Internet of Things

Technology, Communications and Computing

The series Internet of Things - Technologies, Communications and Computing publishes new developments and advances in the various areas of the different facets of the Internet of Things.

The intent is to cover technology (smart devices, wireless sensors, systems), communications (networks and protocols) and computing (theory, middleware and applications) of the Internet of Things, as embedded in the fields of engineering, computer science, life sciences, as well as the methodologies behind them. The series contains monographs, lecture notes and edited volumes in the Internet of Things research and development area, spanning the areas of wireless sensor networks, autonomic networking, network protocol, agent-based computing, artificial intelligence, self organizing systems, multi-sensor data fusion, smart objects, and hybrid intelligent systems.

** Indexing: *Internet of Things* is covered by Scopus and Ei-Compendex **

More information about this series at http://www.springer.com/series/11636

Giancarlo Fortino · Antonio Liotta ·
Raffaele Gravina · Alessandro Longheu
Editors

Data Science and Internet of Things

Research and Applications at the Intersection of DS and IoT

Springer

Editors
Giancarlo Fortino
University of Calabria
Rende (CS), Italy

Raffaele Gravina
University of Calabria
Rende (CS), Italy

Antonio Liotta
Edinburgh Napier University
Edinburgh, UK

Alessandro Longheu
University of Catania
Catania, Italy

ISSN 2199-1073 ISSN 2199-1081 (electronic)
Internet of Things
ISBN 978-3-030-67196-9 ISBN 978-3-030-67197-6 (eBook)
https://doi.org/10.1007/978-3-030-67197-6

This Springer imprint is published by the registered company Springer Nature Switzerland AG
The registered company address is: Gewerbestrasse 11, 6330 Cham, Switzerland

Foreword

The immersive ecosystem of the Internet of Things, with its capillary sensor streams, is flooding information systems, and consequently business processes, with huge amounts of heterogenous real-time raw data.

Such huge data flows truly represent the most precious IoT innovation and at the same time the most complex challenge as Big IoT Data often requires real-time processing and fusion to turn into actionable and smart insights.

Here, Data Science, by means of dedicated methodologies, machine learning techniques, and tools, can unleash the full potential of IoT allowing it to translate raw, fragmented data into meaningful, smart information. The picture that arises is a two-faced coin with IoT representing the ubiquitous data generation infrastructure and Data Science, the first consumer of such data to uncover information and eventually knowledge.

This book, often through the presentation of practical innovative IoT applications spanning from health care to agriculture, investigates how Data Science methods, algorithms, and tools can effectively complement the IoT infrastructure.

Catania, Italy

Min Chen

Preface

In everyday life, the Internet of Things (IoT) paradigm fosters the idea of interconnected entities as computers, humans, objects, devices and sensors, all capable of exchanging data to collaborate and provide unforeseen services, spanning various domains as industrial manufacturing, livelihood resources management (agriculture, water and waste), sustainable and eco-friendly economy, logistics, transportation, smart cities, health care, social networks, nanotechnologies frontiers and others. Like most of the modern technologies, IoT relies on advanced hardware and high-speed networks for its purposes, but these are only the means that endorse the massive underlying stream of data, the real core of IoT.

On the other hand, Data Science (DS) meets IoT with its methodologies, techniques and tools to translate data into information, enabling the effectiveness and usefulness of new services offered by IoT stakeholders. If IoT can indeed be considered the infrastructure of a new world full of amazing possibilities, DS is the key that can really lead to a significant improvement of human life.

It is known, however, that generating data is much easier than extracting something useful from it, even in the presence of high-end performance computer systems. The rate of (even real-time) data creation, especially when considering IoT, is unequivocally higher than the rate of increasing in machine processing capabilities. To complete (and complicate) the picture, the huge amount of data coming from IoT sensors and devices may differ in type, format and semantics, therefore this also demands for a significant effort to compare, discriminate and/or integrate distinct sources of information; researchers tried to condensate all these issues into the term "Big data" that features the evolution of DS.

This book investigates the combination of IoT and DS, specifically how methods, algorithms, and tools from DS can effectively support IoT; the aim is also to present innovative IoT applications as well as ongoing research that exploit modern DS approaches. Readers are offered new issues and challenges in a cross-disciplinary context that involves both IoT and DS fields. In particular, the book spans the following scenarios: (1) smart home architectures; (2) mobile edge computing for health care; (3) centrality measures in complex networks; (4) Low-Power WAN; (5) hydroponic agriculture; (6) collaborative body sensor networks; (7) Unmanned Aerial Vehicles; (8) criminal networks analysis; and (9) indoor navigation systems. These research areas correspond to authored chapters briefly introduced below.

"IoT Aided Smart Home Architecture for Anomaly Detection" by Ibrahim Arif and Nevena Ackovska presents a system for the discovering of unusual events inside homes through the comparison of past versus present behaviors as revealed by data from IoT nodes. The chapter in particular starts reviewing the existing models in the field of anomaly detection in smart homes, then considers fire detection/prediction and fall detection of elderly people as the two most significant anomalies and discusses how the proposed system can effectively detect them and which countermeasures can be carried out to save lives.

"Evolutionary Dynamics and Multiplexity for Mobile Edge Computing in a Healthcare Scenario" by Barbara Attanasio, Alessandro Di Stefano, Aurelio La Corte and Marialisa Scata proposes a framework to conceive cognitive ambient assisted living (AAL) of people with frailty syndromes. The chapter considers how mobile edge computing (MEC) can endorse the shifting of health care from a traditional approach to a distributed patient-centric one, specifically exploiting the cognitive and evolutionary dynamics of complex networks to extract collective knowledge from the AAL of frail people viewed as a social network. In the chapter, the evolution of cooperation between MEC nodes with a low blocking probability is explored to prevent inefficiency in the proposed smart AAL.

"Correlations Among Game of Thieves and Other Centrality Measures in Complex Networks" by Annamaria Ficara, Giacomo Fiumara, Pasquale De Meo and Antonio Liotta investigates how the Game of Thieves (GoT) can be used to compute nodes' and links' centrality in complex network. The GoT is a new algorithm that executes in polylogarithmic time with respect to the state-of-the-art methods that need at least a quadratic time. The chapter illustrates recent literature and discusses several experiments on three correlation metrics in networks differing for nature (artificial and real), type (scale-free, small-world and Erdös-Rényi) and size.

"A LPWAN Case Study for Asset Tracking" by Fabrizio Formosa, Michele Malgeri and Marco Vigo describes a smart application aiming at tracking livestocks using a Low-Power Wide-Area Network (LPWAN), a widely adopted technology within IoT for its low power, long range and low-cost features. The chapter provides the state of the art in industrial IoT platforms useful for tracking applications, then briefly discusses asset tracking, its key features and main use cases, finally introducing the case study together with a discussion on functionalities and performance assessment.

"Implementing an Integrated Internet of Things System (IoT) for Hydroponic Agriculture" by Georgios Georgiadis, Andreas Komninos, Andreas Koskeris and John Garofalakis concerns modern trends in IoT-based agriculture, in particular, how the adoption of Wireless Sensor Networks (WSNs) to measure soil properties, environmental and ambient light conditions enables precision agriculture, allowing the reduction of energy, water, fertilizer and chemicals used for plant growth. After an overview of the current literature, the chapter analyzes the use of IoT in hydroponic agriculture, where plants are placed over a substrate material and continuously watered with nutrient solutions, an innovation leading to very high-quality products although more expensive than other farming approaches.

"A Collaborative BSN-Enabled Architecture for Multi-user Activity Recognition" by Qimeng Li, Raffaele Gravina, Congcong Ma, Weilin Zang, Ye Li and Giancarlo Fortino considers Body Sensor Networks (BSN) used to support the shifting toward human-centric IoT services, and combines them with multi-user Activity Recognition (AR) to analyze the interaction of individuals in a shared environment, a research area in embryonic stage. After a brief discussion on related work on recent multi-user AR, the chapter presents a novel BSN-enabled architecture to support collaborative multi-user AR achieving better responsiveness, latency reduction and higher scalability.

"Collaborative Solutions for Unmanned Aerial Vehicles" by Francisco Fabra, Julio A. Sanguesa, Willian Zamora, Carlos T. Calafate, Juan-Carlos Cano and Pietro Manzoni is a chapter where Unmanned Aerial Vehicles (UAVs), a.k.a. drones, are the main topic. Several issues are outlined, from battery lifetime challenge to privacy, security, and safety, to their social applications, as human transportation and indoor navigation systems, to legal related questions. The chapter then focuses on collaborative solutions for UAVs, joining them into a swarm to carry out a cooperative task; different scenarios are presented, and for each of them a proper protocol to manage all swarm UAVs as a whole is introduced in terms of its finite state machine, together with performance metrics coming from real experiments.

"Graph and Network Theory for the Analysis of Criminal Networks" by Lucia Cavallaro, Ovidiu Bagdasar, Pasquale De Meo, Giacomo Fiumara and Antonio Liotta explores network science tools to investigate criminal networks, also in order to assist law enforcement agencies. To this purpose, the chapter considers the Sicilian Mafia scenario, in particular the set of theoretical tools used to analyze such a criminal network is discussed, the social network analysis is briefly reviewed and finally a study on two real criminal networks related to Sicilian Mafia is presented, showing the dataset extracted from juridical acts, the weights distribution analysis useful to understand suspected interactions and the shortest path analysis that allows to identify trusted affiliates inside the clan who can spread confidential and illegal messages.

"A Data Mining Approach for Indoor Navigation Systems in IoT Scenarios" by Mahbubeh Sattarian, Javad Rezazadeh, Reza Farahbakhsh and Omid Ameri Sianaki considers indoor navigation systems (INS) and discusses the role of data mining techniques in allowing the prediction of what actions INS devices will take, based on past history. After a deep overview on existing related works, the chapter illustrates a navigation model based on the data of a real-world scenario adapted from the emergency ward of a large hospital, and discusses how such previous experiences can be effectively used to suggest better routes; results show improvement both for traveled distance and elapsed time by decreasing the number of turns and obstacles encountered.

Our gratitude is to all chapter contributors, the reviewers and for Mary E. James from Springer for her support and work during the publication process.

Catania, Italy

Giancarlo Fortino
Antonio Liotta
Raffaele Gravina
Alessandro Longheu

Contents

IoT Aided Smart Home Architecture for Anomaly Detection

Ibrahim Arif and Nevena Ackovska

Abstract Internet of Things (IoT), and Information and Communication Technology (ICT) enhance the cities and homes with intelligence using various types of devices that are embedded for sensing purposes. The IoT nodes can collect time-series information for a given purpose, like measuring temperature, air pollution and traffic congestion, motion tracking etc. and provide node behavior and environment interaction. Besides the normal ritual events that happen according the measured parameters, some unusual states, called anomalies, can be detected by using the same measurements. The anomalies that happen at homes, such as fire, elderly people fall etc. are life-critical and their early detection and/or prevention can save lives. In this chapter, we propose an Anomaly Detection System Architecture for smart homes. The fire detection/prediction and fall detection of elderly people are examined as anomalies from the usual, everyday activities. An experimental study, as proof of concept for the fire emergence case, is run by using LSTM neural network architecture. One of the positive features of the proposed methodology is that the faulty readings and false positive alarms on specific parameters are not enough to rise an anomaly detection if those readings are not supported by other parameters.

1 Introduction

As the Industry 4.0 is taking place on the market, the smart city and smart home concepts are getting more popular and becoming everyday phrases for the researchers. The contribution on this field arises as urban population is rising, according the UN reports. As shown in Fig. 1, over 4 billion people already live in urban areas, as is estimated by the end of 2020. Thus, governments invest to improve their cities'

I. Arif (✉) · N. Ackovska
Faculty of Computer Science and Engineering, Ss. Cyril and Methodius University, Skopje, North Macedonia
e-mail: ibrahimarif.cse@gmail.com

N. Ackovska
e-mail: nevena.ackovska@finki.ukim.mk

G. Fortino et al. (eds.), *Data Science and Internet of Things*, Internet of Things,
https://doi.org/10.1007/978-3-030-67197-6_1

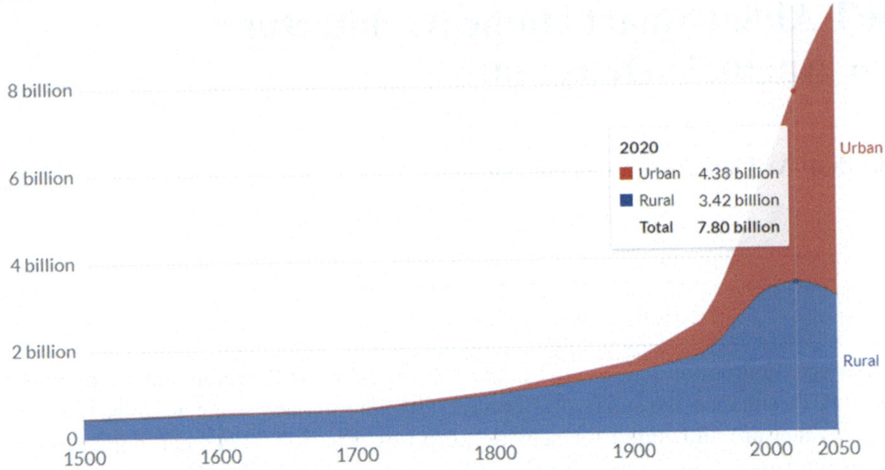

Fig. 1 Urban and rural population projected to 2050: World 1500–2050 [1]

informational infrastructure to provide easier life for their citizens and to adapt to the future technology trends.

With the help of Internet of Things (IoT) and Information and Communication Technology (ICT) the communication and interaction between the devices are provided. Monitoring of the collected data on dashboards is the most common action provided by the software solutions.

Beside monitoring the measured data by including machine learning and other prediction mechanisms, the systems are getting smarter and they take away the human interaction from the measurement and monitoring processes. Since the processed data is enormous and sensing capabilities in IoT devices are precise enough, the smart home systems can detect even parallel events synchronously whilst humans mostly cannot detect them at all.

The common usage of smart city and smart home end-points is to monitor the past or current status of the sensing environment. Moreover, it allows future predictions according the past behavior. The past behavior of the data can be used to devise the normal (usual) behavior of the system. But every system is open to failure or exceptional (unusual) behavior which is presented as anomalous event.

The anomalous behavior of the system can be life critical. Its early detection can save lives in the situations when that particular unusual behavior is dangerous for the people. For example, one such example is the fall of elderly people. The elderly fall anomaly can be presented as most common injury or death cause for elderly people [2].

In this chapter, we are reviewing the existing models in the field of anomaly detection in the smart homes and propose a smart home architecture for anomaly detection of fire and elderly people fall detection cases.

This chapter is organized as follows. Section 2 presents a general survey on the existing models of anomaly detection in IoT used in Smart City and Smart Home.

Section 3 introduces the proposed smart home architecture. Section 4 presents the example application cases for fire emergence and fall detection, where the detailed workflow is shown. Section 5 presents the implementation of an anomaly detection system for the fire emergence case as an experimental study, and the results are reviewed afterwards. Finally, the chapter is concluded with outcome review of the study in the Sect. 6.

2 The Existing Models

In this section, we review the existing models in the application domains of IoT in Smart City and Smart Home where anomaly detection is concerned. The connection between these four concepts and application areas are examined and discussed in sequel.

2.1 Internet of Things

The accessibility of the sensors, communication, and processing units strengthens the power of IoT to be used in everyday devices. Since the main goal of IoT is the devices to be connected to the internet and to be accessed from anywhere, it provides a wide area of application and an open window for scientific research. As the price of the IoT nodes has become cheaper, nowadays, it is easier to build a smart environment in comparison with the previous years. Other recent scientific reports also discuss that cheap and useful hardware architecture and solutions are existent.

The authors in [3] categorize the IoT components into data acquisition system, data communication system, and data processing system. They emphasize that the data communication system commonly consists of standardized platforms and protocols besides the data acquisition and processing systems which are project dependent. Furthermore, they classify the IoT as; (i) consumer IoT (CIoT) where the systems occupy the customer's space and the data is sent to centralized server or an individual gadget for processing, and (ii) industrial IoT (IioT) where technologies like big data, machine learning, device-to-device communication, automation, machine-to-machine (M2M) communication etc. are emerged to perform different functionalities.

The security aspect is mostly neglected, but it represents a big challenge for IoT devices since everything is connected to the internet and can be accessed by its IP. The authors in [4] depict the design-time and runtime approaches to handle safety and security in IoT. The study also consists of literature review on natural experiments in safety and security. On the other hand, in another study [5] a prototype of a reliable, cost effective and simple to use device for security testing of smart grid end point devices is proposed and evaluated.

2.2 Smart Cities

Thanks to the IoT and ICT the cities are getting more intelligent. By sensing the city and acting accordingly the sensed information enables the city to control itself with or without human interaction. The smart city concept covers a wide area of topics, such as energy sources, communications, traffic management, smart health, smart environment, smart home etc. as depicted in Fig. 2.

The smart city can be developed by using 2 approaches: (i) top-down approach where the government standardizes the architectures and centralizes the development and information flow, and (ii) bottom-up approach where every end point builds its own smart environment and makes it eligible for further integration into the smart city.

The authors in [6] present an Intelligent Unmanned Aerial Vehicle (UAV) for urban surveillance case studies. They also develop an intelligent IoT platform for UAV which integrates the 4 aspects: (i) Data, (ii) Connectivity, (iii) Device, and (iv) Service, which provides mobile platform for multiple services.

Energy is one of the most important elements in the smart cities. The study [7] represents an applied IoT architecture for smart microgrids. Its two main design goals are: (i) to go beyond basic electricity services with purpose of using locally produced renewable energy, creating flexible exchange for the energy resources and

Fig. 2 Smart city

contribution realization of the large-scale smart grid, and (ii) to meet high level requirements like being flexible, scalable, reliable, and secure to ensure integration and interoperability with other microgrids.

The IoT middleware architecture for smart cities, which provides communication between heterogeneous systems in the city, is discussed in [3]. The need of IoT middleware is taken in consideration from 5 points of view: (i) Autonomous city, (ii) Centralized control for authorities, (iii) Reduced capex and opex, (iv) Sustainability and maintainability, and (v) Vendor neutrality.

In [8], the authors present a mobile (drive-by) sensing platform where sensing devices are embedded to scheduled vehicles like busses, trash trucks etc. and unscheduled vehicles like taxis. The architecture consists of 3 layers: (i) Sensing layer, (ii) Cloud layer and (iii) Application layer. They overview the sensor types in 5 classes: (i) Ambient fluid properties, (ii) Electromagnetic properties, (iii) Urban envelope properties, (iv) Photonic properties and (v) Acoustic properties, and discuss their potential applications in their platform.

The study in [9] covers most of the branches depicted in Fig. 2 like smart home, smart parking, surveillance, environmental pollution, vehicular traffic, and weather and water system. They define a smart city architecture for big data analytics which consists of 4 tiers: (i) Bottom Tier (Data generation and collection tier), (i) Intermediate Tier-I (Communicator tier), (iii) Intermediate Tier-II (Data management and processing tier), and (iv) Top-Tier (Data interpretation tier).

2.3 Smart Homes

As a branch of smart cities (Fig. 2), smart homes represent a wide area of research since it is the first main terminal of bottom-up approach which nowadays is more convenient to be applied.

Currently, almost all gadgets in the house are smart enough thanks to the IoT and ICT. From air conditioning to entertainment devices (Fig. 3), everything is becoming connected to the internet, thus, the user is able to control them from anywhere.

The features of smart homes such as reducing human effort, helping old and handicapped people, and saving electricity are presented in [10]. On the other hand, the authors in [11] describe the smart home as 'IT enabled home that provides comfort, convenience, security along with energy efficiency thorough prediction, analysis of data and automation of devices."

As a different approach of developing a complete framework, the authors in [12] present three layered *Cloud-assisted Agent-based Smart home Environment (CASE)* architecture which relies on distributed multi-agent paradigm and cloud technologies. Their objective is to recognise activities of a person in a home environment by using various sensor measurements.

A comprehensive review on the literature of IoT for smart homes is presented in a study [13] where practical design challenges for data processing are discussed.

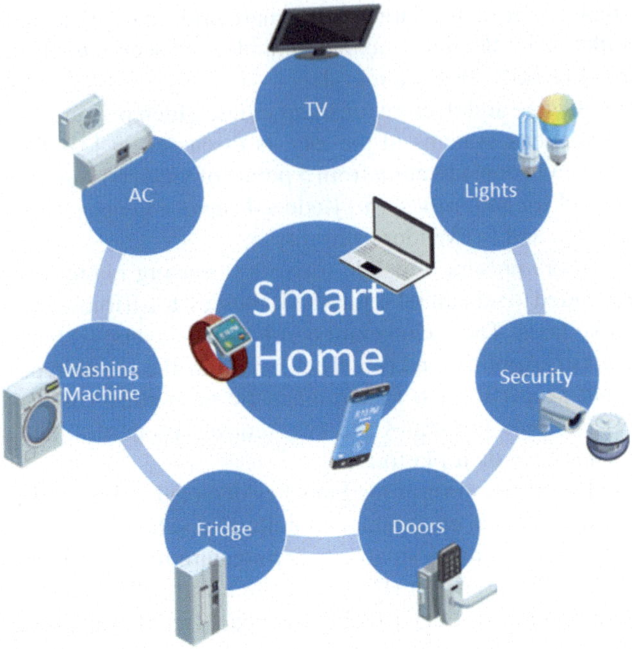

Fig. 3 Smart home

2.4 *Anomaly Detection Systems*

Unusual behavior or failure of a system is defined as an anomaly. It can be system dependent like failure of a system or its sub-system, or system independent where the anomaly can occur by external factors, e.g. fire.

A novel approach on the anomaly detection by using neural network Self-Organizing Maps is presented in [14]. The authors classify anomalies as: (i) unusual activities where the tracked user is at unusual location at home for a focused time (e.g. sitting in kitchen at bed time), (ii) unusual long activities where the activities last longer than expected due to falls or loss of consciousness, and, (iii) unusual short activities where the activities last shorter than expected which can mean the user is facing health problems (e.g. waking up at night hours). On experiments, their system predicts the anomalies with accuracy of 83.8% to 99.1% for unusual short activities and unusual activities respectively.

In [15], the authors examine the anomaly detection attacks on smart electricity meters. Vector, Honesty coefficient, and KLD based anomaly detection algorithms are presented for identifying compromised meters in Advanced Metering Infrastructure with the concept of Swarm Intelligence (algorithms inspired from nature like ant colonies, flock of birds, fish schools etc.). The detection probability of the presented system is close to 1.0 whereas the false alarm is 0.17.

The authors in [16] depict that threshold-based algorithms have been used in most of fall detection area, followed by decision tree based algorithms, genetic programming, and support vector machines. According to the used sensor types, the applied fall detection algorithms also vary.

Besides the anomaly detection/prediction accuracy, the literature concludes that lowering the false alarm is also important on the anomaly detection area.

3 Proposed Smart Home Architecture

Considering the aforementioned concepts and architectures we have developed a smart home architecture based on the architecture in [8, 11], for anomaly detection integration as depicted in Fig. 4. The architecture consists of 3 layers. They will be explained in the following subsections. Examples for using this architecture are given in Sect. 4.

3.1 *Layer I - Sensing Layer*

The sensing layer covers all smart home sensors which sense the environment. Every sensor has its own sensing mechanism and converts the measured value to a meaningful parameter. This layer also includes M2M (Machine to Machine) communication where the IoT devices communicate with each other through wireless protocols like ZigBee, Xbee, Wifi, Bluetooth etc.

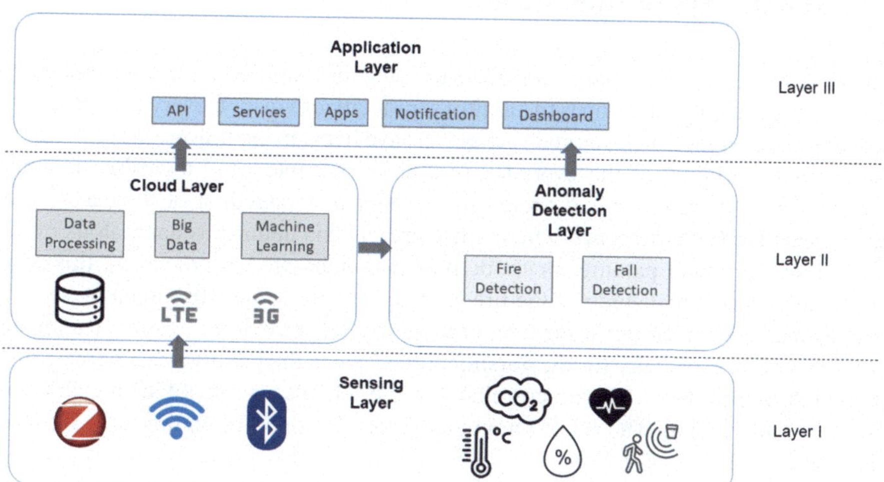

Fig. 4 Proposed smart home architecture

3.2 Layer II - Cloud Layer and Anomaly Detection Layer

The cloud layer provides network and communication over LTE, GSM, 2G/3G, and storage capabilities. Moreover, since the computational power is higher in this layer, machine learning, big data and other data processing technologies can be applied for better knowledge acquisition and event predictions.

The anomaly detection layer comes into play beside the cloud layer where an anomaly situation is being detected according to the processed data. The anomaly is detected when the time-series data is exceptional to its normal behavior and it is mostly an outlier to the statistical data. For example, if an instant rise in room temperature over the expected normal temperature change occurs, there is a possibility to a fire being the cause of the abnormal heating. Depending on the anomaly domain, this layer is supposed to raise an error and provide interaction with the corresponding IoT nodes from Layer I according to the use cases or alert the user over Layer III applications.

3.3 Layer III - Application Layer

The application represents a general overview for the user of the sensing devices. According to the implemented project, the applications can vary. In our case, for anomaly detection, the notification mechanism plays a vital role, since the user can respond to the anomaly as soon as it occurs.

4 Example Application Cases

The work presented in this paper considers fire detection and fall detection as anomaly cases. Since the final intention is to monitor senior citizens in their everyday activities, the early detection of such anomaly cases can save lives. In the following subsections the normal workflow of the everyday routine is presented and then the anomaly occurrence on these cases is depicted for a better understanding of the cases. The first case is for fire emergence where a fire presence is defined as anomalous event in a house. Particular parameters are defined and elaborated as crucial for this case (4.1). An experimental study of the fire emergence case is then discussed in Sect. 5. The second case is for detecting fall, particularly fall of elderly people. This is an important case, especially for the nursing homes, since the fatalities that happen due to the fall of elderly people are correlated to the late discoveries of such cases and thus late intervention. The set of parameters that are important to this case are also elaborated (4.2).

4.1 Fire Emergence Case

Nowadays, almost every house in the developed countries has at least one IoT node. The very common fire detection mechanism is smoke detection system where an alarm goes off when smoke is detected or watering mechanism is activated to extinguish the fire. According to the sensitivity they can be activated even by cigarette smoke or a smoke which can be detected through window. To strengthen the fire decision, the ambient temperature can be taken into account.

Refrigerators, air conditioners, heating devices etc. have sensors for measuring the ambient temperature of the environment. By processing the measured temperature as time-series the normal behavior of the system can be extracted. Thus, when exceptional data pattern occurs, e.g. instant temperature increase, the event is considered as anomaly and it is treated as fire.

For the related case, the indoor station for measuring air pollution which is presented in a study [17] is a practical fit. Although it is built for measuring air pollution, its sensing on temperature, humidity, and CO2 can provide reliable data for integration with fire detection as anomaly in the environment.

Figure 5 shows the fire detection workflow where the time-series data is proceeded to the prediction model, then to the anomaly detection model where an anomaly is detected. The unusual behavior of the sensors' time-series data is summed up as a weighted error where the weights represent importance of the sensor for the related anomaly, e.g. temperature data is more important than humidity data, and if the sum is greater than a defined threshold the system alerts an anomaly. The experimental study for this case is presented in Sect. 5.

In response to the fire detection, for an ideal smart home example, the system should notify the corresponding user and emergency teams. Moreover, the system should communicate with other smart IoT devices in order to force them to work in crisis-mode where a bigger disaster can be prevented, e.g. autonomously disabling the natural gas infrastructure, activate the anti-fire system etc.

The early detection and prevention of fire is important for human lives since the fires are mostly detected when they already become dangerous. Thus, the sooner the emergency forces are informed the faster they respond to the disaster.

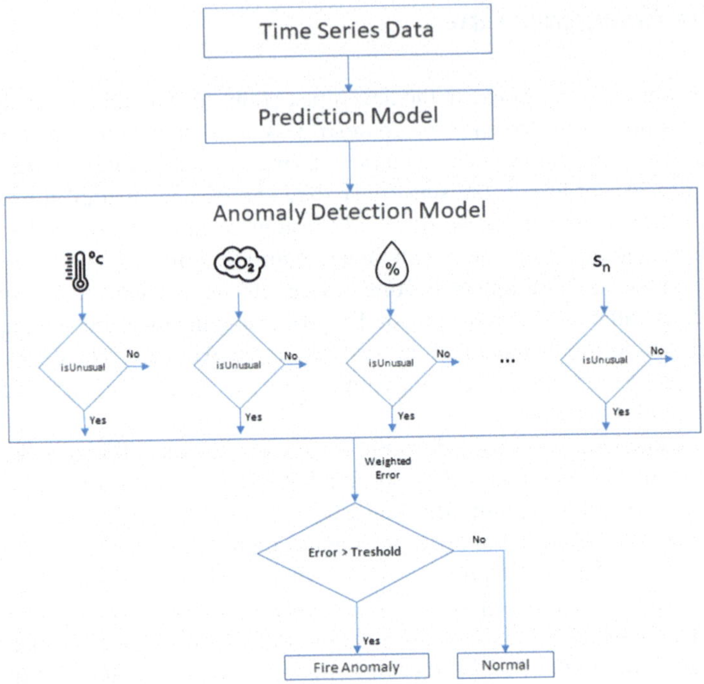

Fig. 5 Fire detection workflow

4.2 *Fall Detection Case*

The most common injury that can lead to death in elderly people is their fall [2]. Especially, being alone calling for help while there is no one around is great problem in their age.

The authors in[18] present non-invasive environmental sensors for detecting user activities inside the house. Furthermore, the papers [16, 19], among others, present fall detection systems that contribute to this case. In [14], an experimental study on fall detection as anomaly is presented.

The elderly people, mostly in rehabilitation centers, are tracked by motion sensors rather than cameras in order to not disturb their privacy. By using the motion sensors' past data and current data as time-series, the anomaly detection system can learn the usual behavior of the related elderly user. In case of abnormal situation, the system detects the unusual activity, e.g. the elderly fall in our case, and raises an alarm to the related units.

Another approach is to use attached IoT device to the corresponding user which measures heart rate, temperature etc. parameters of the elderly people. This information creates a user profile for its usual activities. In case of fall or conciseness loss the system detects the anomaly and reports to the related units.

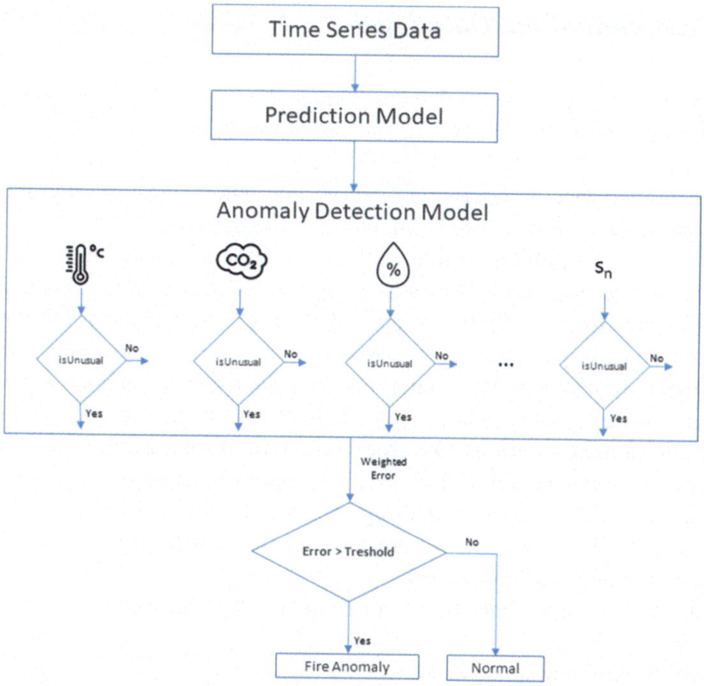

Fig. 6 Fall detection workflow

Figure 6 shows the fall detection workflow. The anomaly detection model calculates a weighted error for anomalous behavior of the related data and then the system decides whether the activity is anomalous or not. Most elderly fall deaths are based on late detection of the falls. These systems can provide early detection and notifying mechanism for fast interventions which may save lives.

5 Experimental Study

As a proof of concept of the proposed architecture in Sect. 3, we have implemented an experimental anomaly detection system for the fire emergence case (4.1). The observation study follows the workflow shown on Fig. 5 and covers all of the aforementioned aspects for detecting an anomaly.

The implementation is inspired from a comprehensive study [20] on short-long term anomaly detection in heterogeneous sensor networks. Detecting faulty readings caused from the sensor itself and anomalous behaviour of the system is the main focus of this study. This is also included in our experimental study where the data set consists of faulty readings besides the anomalous readings.

5.1 Description of the Data Used

In this experiment, we focus on detecting anomalies by looking on heterogeneous parameters, thus, the main factor of choosing appropriate data set for the experimental study is the data set to include time-series measurement from various sensors. For the fire emergence case; temperature, carbon monoxide, and humidity are some of the useful parameters and should be included in the data set.

There are data sets publicly available that contain many of the measured parameters that are vital to our study. The data set used in this research is [21] published on Kaggle by Home Office AirPi. It is a very rich data set which consists of temperature, relative humidity, air quality, carbon monoxide, nitrogen dioxide among others. The measurements are taken in 5 min intervals. The data set was originally created for monitoring of the air quality in home environment, and eventually to asses the impact of changeable home conditions like entering/leaving room, running heating device in the room, even experiment a fire case. The main objective of the related project was to examine the data when something occurs in the related system.

Oppositely, in this research, we are looking through the data set to detect the environmental changes as anomalies.

The data set [21] represents 14,572 readings for 9 parameters:

Volume [mV]: Ambient noise level.
Temperature-DHT [Celsius]: Temperature reading from the DHT22 sensor in degrees Celsius. *Has a small number of faulty readings due to sensor failure.*
Pressure [Hectopascal]: Air pressure from the BMP085 sensor in hectopascal.
Temperature-BMP [Celsius]: Temperature reading from the BMP085 sensor in degrees Celsius.
Relative_Humidity [%]: Relative humidity reading from the DHT22 sensor. *Has a small number of faulty readings due to sensor failure.*
Air_Quality [Ohms]: Air quality measurements from the TGS 2600 sensor.
Carbon_Monoxide [Ohms]: Carbon monoxide measurements from MiCS-5525 sensor.
Nitrogen_Dioxide [Ohms]: Nitrogen dioxide measurements from MiCS-2710 sensor.
Because the initial time-series readings are non-anomalous [22] the data set is divided into 10% of training set and 90% of test set, as seen in Fig. 7. The test set is our point of interest information, where faulty sensor readings, and instant changes can be detected as anomalies.

5.2 Implementation Using Long Short-Term Memory (LSTM)

Every measurement in time-series data is related to the previous readings on the sequence, and predicting the upcoming data point is a big challenge for building an intelligent system. Since the Recurrent Neural Networks (RNN) are basically designed to deal with sequential data, time-series in our case, we have decided to use the capabilities of RNN for implementation phase of the experimental study. Furthermore, Long Short-Term Memory (LSTM) is a powerful RNN architecture which deals with relational patterns on long sequences with unknown lengths [22].

In this experimental study, we are using the well known machine learning library, Keras, to implement LSTM based anomaly detection model. Firstly, we are pre-processing the data into 10% training and 90% testing data, as mentioned before. For training, alongside each data point x_t and its next reading x_{t+1}, where t is the current state, a sequence of previous readings $x_{t-n}, \ldots, x_{t-3}, x_{t-2}, x_{t-1}$, where n is *unroll length* and in our case is set to 50, is proceeded for training the LSTM. On the other hand, the same method is used for testing, but this time the next reading x_{t+1} is predicted by the trained LSTM model.

Secondly, after defining the training and testing data, we are setting the LSTM layers and tuning the necessary parameters in order to get acceptable training results. It is worth mentioning that 10% of the training data is set as *validation data* to validate the training performance of the prediction model. The Fig. 8 represents the resulting performance of the learning phase. After running tests on parameters with default of 50 epochs, we have discovered that some of them (Temperature-DHT, Temperature-BMP, Carbon_Monoxide, Nitrogen_Dioxide, Pressure) request less epochs to effectively train the prediction model whilst readings for Light_Level request more epochs for better fit. As a result, by looking at *train loss* and *validation loss* graphs in Fig. 8, almost all parameters fit with high success rates except Air_Quality and Volume sequences.

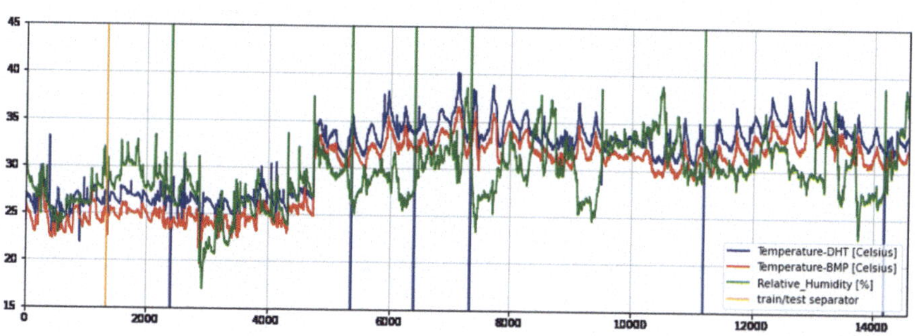

Fig. 7 Look up to sensor readings for temperature (2 different sensors) and humidity

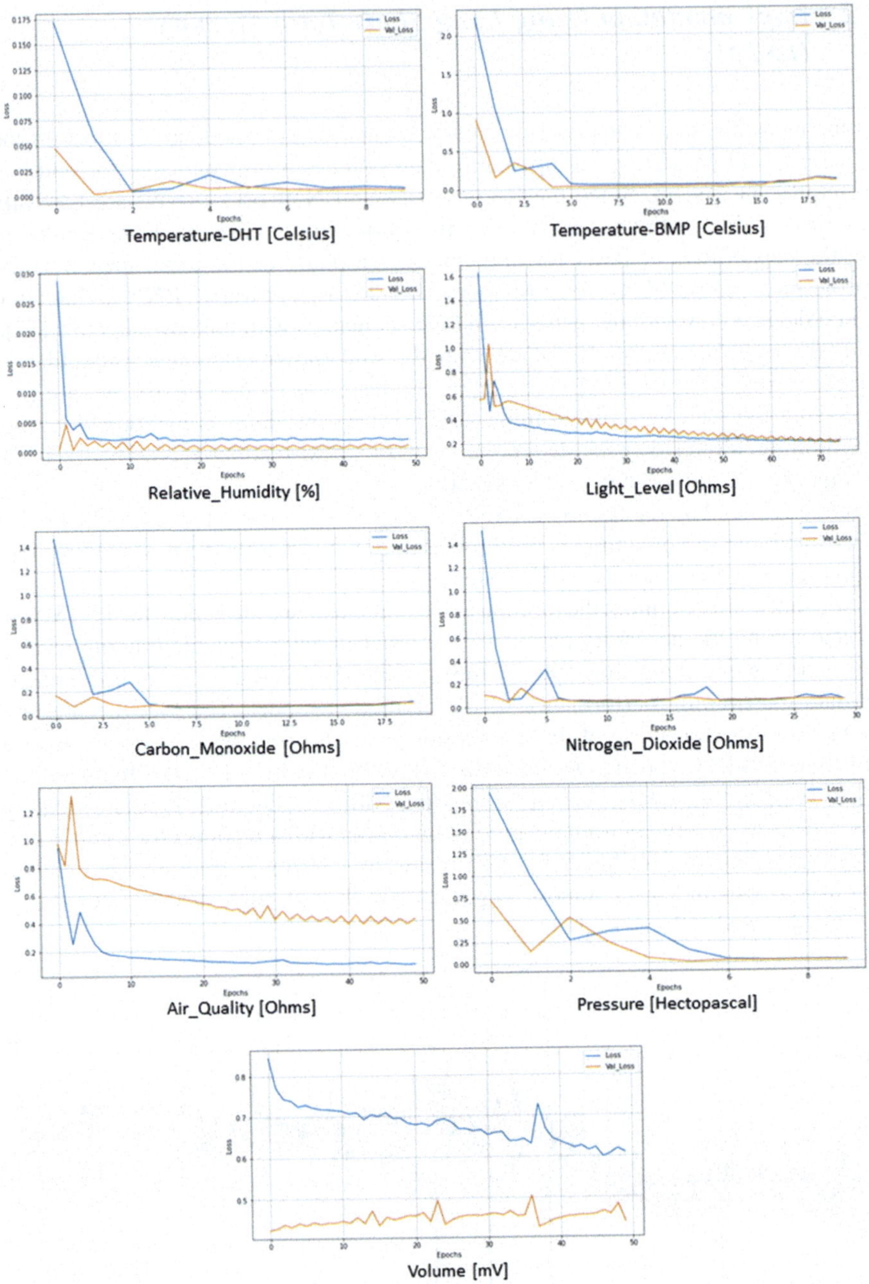

Fig. 8 Learning performance: training loss and validation loss evaluation over epochs for each parameter

Finally, the predicted data is compared with the real data in Fig. 9. The reason for training the Air_Quality and Volume data with higher loss can be inspected from the related graphs. Furthermore, the results are afterwards examined to detect anomalies.

5.3 *Experiment Results*

To detect the anomalies on the test data, we are extracting the differences for every point for each parameter. $d_t = |p_t - r_t|$ where the difference d_t is the absolute difference between the predicted value p_t and the real value r_t at a given time t. Thus, a set of differences d is constructed for every parameter.

Afterwards, we set the anomaly factor to 1% ($anomaly\_factor = 0.01$) which defines that the largest 1% differences from the set d are selected as anomalies. The Fig. 10 presents the anomalous points for each parameter.

To discuss, by looking at the graphics for each parameter, the *last readings of the first quarter* and most of the *last quarter* include anomalous data. The most important outcome of the implementation is that the anomalies are detected not only on a single, but on multiple parameters. Thus, we can conclude that a detection of unusual event on a single parameter is assisted by various parameter readings. Therefore, false positive situations can be detected when a specific parameter alarms a problem and others do not support that.

In addition, the readings from sensor failure also appear as anomalies and it is a big issue to determine whether it is a real world anomaly or not. As mentioned before, this situation is also solved by registering unsupported faulty readings as false positive alarms.

Finally, although we have worked on offline data, the experimental study results are promising and bring the confidence that the system will be successful on streaming/real-time data because the previous sequence is known and the predicted data will be measured on the next reading.

6 Conclusion

In this chapter we present an architecture that integrates the concepts of a smart home architecture with the anomaly detection subsystem. This subsystem is capable of detecting the unusual behavior and then reporting it to the other, more traditional layers of the architecture. Integrating an Anomaly Detection module to an existing IoT infrastructure can add enhanced smartness to the existing smart home system by using the already measured values such as temperature, light etc.

Additionally, two application cases are elaborated in more detail, fire detection anomaly and fall anomaly. The early detection of fall in elderly people can save lives, since the late detection and intervention are the main reasons of them not surviving a fall. The fall can be detected as an anomaly of the normal person's routine. In this case, the notifying mechanism plays a vital role in order to save the life of the elderly.

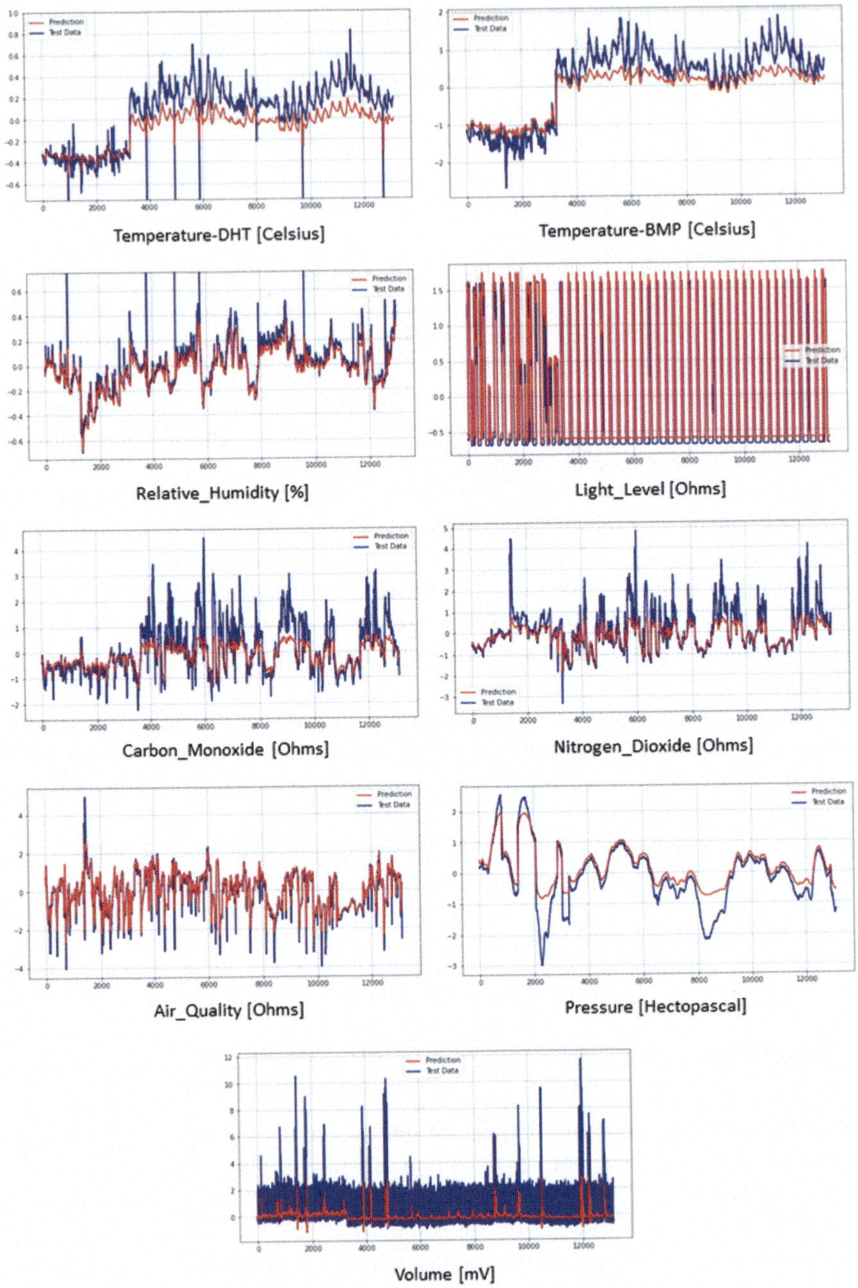

Fig. 9 Differences between prediction and test data, for each parameter

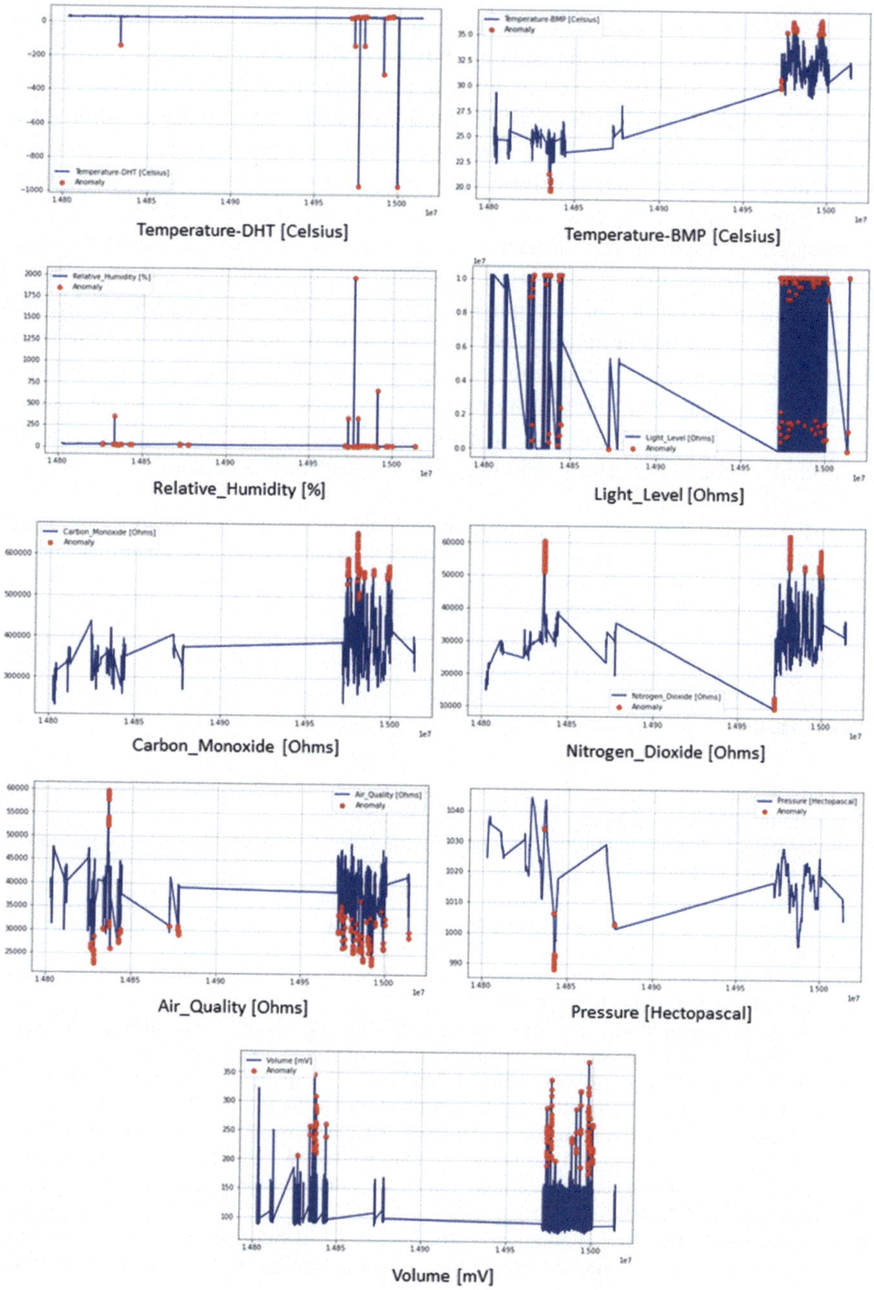

Fig. 10 Pointing the anomalous readings for each parameter

As explained in the fire case, the early stage of a fire can be detected by using existing temperature measuring sensors, but also other types of sensors which are not primarily intended for fire detection, but can relate to a fire. After detecting the fire as an abnormal activity, the notifying actions play a vital role for informing the related units for emergency. The communication with other IoT devices can also work against the fire anomaly, by forcing the other IoT nodes to work in crisis-mode.

The experimental data set was perfect fit to achieve the expected outcome from this research. The synchronous detection of an anomaly by various parameters shows the power of the system in pointing the unusual events with high precision. Some detected outliers on particular parameter are critical to the system. In addition, besides detecting the real world anomalies, we are extracting the faulty readings from some sensors too, which is a big challenge to classify them as a sensor failure instead of fire occurrence. Therefore, the faulty readings and false positive alarms on specific parameters are not enough to rise an anomaly detection if those readings are not supported by other parameters, which is also result from this research.

Acknowledgements This work was partially financed by the Faculty of Computer Science and Engineering at the Ss. Cyril and Methodius University in Skopje, North Macedonia. The authors would like to thank Angela Momirovska and Aynur Kaso, from Ss. Cyril and Methodius in Skopje, North Macedonia, for their assistance.

References

1. "Our world in data." [Online]. https://ourworldindata.org/grapher/urban-and-rural-population-2050. Cited 30 Apr 2020
2. Rubenstein, L.Z.: Falls in older people: epidemiology, risk factors and strategies for prevention. Age Ageing **35**(2), ii37–ii41 (2006)
3. Joseph, T., Jenu, R., Assis, A.K., Kumar, V.S., Sasi, P., Alexander, G.: Iot middleware for smart city:(an integrated and centrally managed iot middleware for smart city). In: IEEE Region 10 Symposium (TENSYMP), pp. 1–5. IEEE (2017)
4. Wolf, M., Serpanos, D.: Safety and security in cyber-physical systems and internet-of-things systems. Proc. IEEE **106**(1), 9–20 (2017)
5. Saleh, M., Al Barghuthi, N.B., Alawadhi, K., Sallal, F., Ferrah, A.: Streamlining "smart grid end point devices" vulnerability testing using single board computer. 2018 Advances in Science and Engineering Technology International Conferences (ASET), pp. 1–6. IEEE (2018)
6. Giyenko, A., Im Cho, Y.: Intelligent uav in smart cities using iot. In: 2016 16th International Conference on Control, Automation and Systems (ICCAS), pp. 207–210. IEEE (2016)
7. González, R.M., Wattjes, F.D., Gibescu, M., Vermeiden, W., Slootweg, J.G., Kling, W.L.: Applied internet of things architecture to unlock the value of smart microgrids. IEEE Internet Things J. **5**(6), 5326–5336 (2018)
8. Anjomshoaa, A., Duarte, F., Rennings, D., Matarazzo, T.J., deSouza, P., Ratti, C.: City scanner: building and scheduling a mobile sensing platform for smart city services. IEEE Internet Things J. **5**(6), 4567–4579 (2018)
9. Yadav, P., Vishwakarma, S.: Application of internet of things and big data towards a smart city. In: 2018 3rd International Conference On Internet of Things: Smart Innovation and Usages (IoT-SIU), pp. 1–5. IEEE (2018)
10. Vaidya, V.D., Vishwakarma, P.: A comparative analysis on smart home system to control, monitor and secure home, based on technologies like gsm, iot, bluetooth and pic microcon-

troller with zigbee modulation. In: 2018 International Conference on Smart City and Emerging Technology (ICSCET), pp. 1–4. IEEE (2018)
11. Baig, M.N., Himarish, M.N., Pranaya, Y., Ahmed, M.R.: Cognitive architecture based smart homes for smart cities. In: 2018 2nd International Conference on Trends in Electronics and Informatics (ICOEI), pp. 461–465. IEEE (2018)
12. Cicirelli, F., Fortino, G., Giordano, A., Guerrieri, A., Spezzano, G., Vinci, A.: On the design of smart homes: a framework for activity recognition in home environment. J. Med. Syst. **40**(9), 200:1–200:17 (2016)
13. Stojkoska, B.L.R., Trivodaliev, K.V.: A review of internet of things for smart home: challenges and solutions. J. Clean. Prod. **140**, 1454–1464 (2017)
14. Novák, M., Jakab, F., Lain, L.: Anomaly detection in user daily patterns in smart-home environment. J. Sel. Areas Health Inform **3**(6), 1–11 (2013)
15. Paikrao, P.S., Bose, R.: Anomaly detection algorithms for smart metering using swarm intelligence. In: Proceedings of the 1st International Workshop on Future Industrial Communication Networks, pp. 3–8 (2018)
16. Ahmed, M., Mehmood, N., Nadeem, A., Mehmood, A., Rizwan, K.: Fall detection system for the elderly based on the classification of shimmer sensor prototype data. Healthc. Inform. Res. **23**(3), 147–158 (2017)
17. Ilievski, A., Dojchinovski, D., Ackovska, N., Kirandziska, V.: The application of an air pollution measuring system built for home living. In: International Conference on Telecommunications, pp. 75–89. Springer (2018)
18. Dimitrievski, A., Zdravevski, E., Lameski, P., Trajkovik, V.: Towards application of non-invasive environmental sensors for risks and activity detection. In: 2016 IEEE 12th International Conference on Intelligent Computer Communication and Processing (ICCP), pp. 27–33. IEEE (2016)
19. Yacchirema, D., de Puga, J.S., Palau, C., Esteve, M.: Fall detection system for elderly people using iot and big data. Procedia Comput. Sci. **130**, 603–610 (2018)
20. Cauteruccio, F., Fortino, G., Guerrieri, A., Liotta, A., Mocanu, D.C., Perra, C., Terracina, G., Torres Vega, M.: Short-long term anomaly detection in wireless sensor networks based on machine learning and multi-parameterized edit distance. Inf. Fusion **52**, 13–30 (2019)
21. Home Office AirPi, "Indoor climate measurements using an AirPi: A Raspberry Pi shield kit.", Kaggle, [Online]. https://www.kaggle.com/mvolkerts/home-office-airpi. Cited 09 Aug 2020
22. Malhotra, P., Vig, L., Shroff, G., Agarwal, P.: Long short term memory networks for anomaly detection in time series. In: Proceedings, vol. 89, pp. 89–94. Presses universitaires de Louvain (2015)

Evolutionary Dynamics and Multiplexity for Mobile Edge Computing in a Healthcare Scenario

Barbara Attanasio, Alessandro Di Stefano, Aurelio La Corte, and Marialisa Scatá

Abstract Today, on the wake of the evolution of 5G towards 6G, the complex joining of existing communication systems and the socio-technical aspects of human interactions, acquire a growing scientific interest. In particular, this can become decisive to establish innovation in healthcare. Complex networks theory and evolutionary dynamics can play a key role in designing a smart healthcare system, and enable to provide statistical estimators to understand which measures and how would be needed to dynamically manage requirements and needs. Thus, following this approach, we propose a framework for a smart healthcare scenario, to design a cognitive ambient assisted living of frail people connected. We consider the multiplex networks to represent two interdependent networks, the mobile edge computing nodes network and the social network of frail people. In the case of a multi-services environment, we evaluate the impact of evolutionary dynamics of cooperation of mobile edge computing nodes in the system. Our findings show how the evolutionary dynamics of mobile edge computing nodes allow decreasing the blocking probability, with the increasing of cooperators in the considered scenario.

B. Attanasio (✉) · A. La Corte · M. Scatá
Dipartimento di Ingegneria Elettrica, Elettronica e Informatica,
Università degli Studi di Catania, v.le A. Doria 6, 95125 Catania, Italy
e-mail: barbara.attanasio@phd.unict.it

A. La Corte
e-mail: aurelio.lacorte@unict.it

M. Scatá
e-mail: lisa.scata@dieei.unict.it

A. Di Stefano
School of Computing, Engineering & Digital Technologies, Teesside University,
Middlesbrough TS1 3BX, UK
e-mail: distefano.alessandro@gmail.com

G. Fortino et al. (eds.), *Data Science and Internet of Things*, Internet of Things,
https://doi.org/10.1007/978-3-030-67197-6_2

1 Introduction

The ongoing process of telecommunications evolution within 5G networks, and the recent development towards the 6G along with the increasing number of users and consequently the demands of growing bandwidth-intensive services and applications, is creating a large volume of traffic, making a compelling and complex dare that still lies ahead [20, 32]. Some specific attributes used to characterize networks such as interference, coverage, throughput, robustness and cost, are not able to describe the crucial aspects of the organization in terms of dynamics and structural degree of future wireless and mobile networks. For this reason, 5G towards 6G, will be focused on complex networks topics such as densification, heterogeneity and distributed and self-organizing decision-making [56]. Moreover the role of 5G and the future innovations of 6G will allow to turn into heterogeneous, cognitive and constantly-changing connected systems of things (Internet of Things) and people (Internet of People) [30, 41, 52]. Over recent years, the interest on some sociotechnical design aspects, that take into account the psychological, collective dynamics, in addition to legal and economic requirements, is growing so fast. This approach, that will be an important new point for 6G [20], would also facilitate the development of new systems and applications, trying to quantify how their introduction impacts on personal life conditions, community, and societal levels, particularly on the design of healthcare systems.

In the modern information society, given that telecommunication networks represent one of the largest scale factor with renovations occurring continuously over the time, it is crucial to reason about the trajectory of the future applications, such as healthcare[3, 17, 52]. Healthcare is experiencing a rapid transformation from traditional approach to a distributed patient-centric one, representing also a significant aspect in the new economy [3]. There is a challenge in public healthcare sector also due to the lack of adequate facilities that are driving the way in which health services are delivered to patients. Smart homes, smart city, which integrate health and other technologies, lead to design innovative cognitive and dynamical environment in which healthcare services are being provided to patients. An interesting scenario can be represented by an Ambient Assisted Living (AAL) in which it is provided the assistance of people with recognized frailty syndromes [51].

Frailty is a vulnerability condition to stressors that results in an increased risk of adverse healthcare outcomes [51]. In this context, the evolution of 5G and the implementation of Internet of Things (IoT) and Internet of People (IoP) have encouraged the development of many services able to collect data from patients, to study the effectiveness of treatments for patients and frail people. There is a lack of features which insert into the systems the capability of spotting collective behaviors, envisioning dynamics of communication networks, that represent a key aspect to learn on how properly monitor them. Thus, to do this and at the same time to collect data about their habits, social interactions and health without obstructing their lives, smart devices should be wearable and wireless. These devices, sensors and actuators are the leading character of the so called "Internet of Healthcare Things" [37] and they are characterized by constrained energy, memory and processing capacity. Moreover, it could be useful to provide computing ability in near proximity of end users, through mobile edge computing (MEC) nodes.

In order to plan suitable healthcare systems that exploit the communication technologies, the complexity of social networks, and the knowledge from data and behavioral collective dynamics [58, 59], this paper is focused on proposing an innovative approach, based on cognitive and evolutionary dynamics and complex networks applied to a scenario represented as a smart AAL which provides healthcare multi-services to frail people. Through the multiplex networks representation we envision the AAL of frail people as a social network able to produce collective knowledge through dynamics and big data. Patients, frail people are represented as nodes and they are linked to each other through different ties. In addition, we represent the MEC as a service-based multiplex networks which depends on knowledge and data from social network of frail people. Data are related to habits, clinical conditions, behaviors, collective awareness, and service requests.

In this paper, we propose a framework which relies on the representations through multiplex networks to design a cognitive dynamical multi-services environment for smart healthcare into AAL of frail people. We explore the evolution of cooperation between MEC nodes with a low blocking probability, in order to prevent inefficiency in the proposed smart AAL. To this aim, we model MEC servers as nodes of a multiplex network, layered as service-based, and evaluate the evolution of cooperation among them thanks to a game-theoretic approach. An interesting point to untangle, introducing the multiplex representation of users and MEC, is the knowledge mining of the network structure and its dynamics. The analysis of coexistence of various types of interactions among users on a complex network [9], allows us to unveil hidden behaviors that this kind of network can exhibit. Moreover, the Evolutionary Game Theory (EGT), which constitutes the mathematical framework to study the evolution of strategic interactions within a complex system, highlights the interdependence between the strategy of a node and the others which are connected to it in the different layers and from which depends the evolution of collective dynamics.

The paper is organized as follows: after an introduction, in Sect. 2 we discuss background and methods pointing out, in the order, the main features of the smart healthcare scenario, the MEC, the multiplex networks and the EGT. In Sect. 3 we show the proposed model and framework. In Sect. 4 we discuss results and findings showing the simulations conducted. Finally, we present conclusion and future works in Sect. 5.

2 Background and Methods

2.1 Internet of People in an Opportunistic Network Scenario

The fast expansion of the Internet at the edge, and the tighter interactions between human users and their personal mobile devices push toward an Internet where human beings become more central than ever. Humans with their personal devices act as their proxies in the cyber world and they are fundamental tools to sense the physical world, leading to the so-called Cyber–Physical convergence [19]. The vast availability of personal devices (e.g., smartphones) carried by humans, and IoT devices (e.g., sensors) embedded in almost every physical object, together with the pervasiveness of wireless communication technologies and the massive amounts of data describing processes, phenomena and behaviors of the physical world are crucial in this phenomenon. Data from the physical world are analyzed in the cyber world to generate appropriate representations of the physical entities and phenomena. Based on this knowledge, services not only interact with each other, but also control the status of the physical world itself, through actuators [46]. In addition, to collect information from the cyber world around us, end devices which move with us and act in the physical world, leave digital traces of our behaviour in the virtual world as: our mobility patterns, our social relationships, our opinions, etc. All this incentivise a radically new Internet paradigm, the IoP [19, 46], where human beings and their personal devices are not seen merely as end users of applications, but become active elements of the Internet. This paradigm is based on the observation that, since the edge of the Internet is primarily populated by human personal devices, or by devices embedded in physical objects with which humans constantly interact, the Internet network paradigms should take into account human behaviours in the design of all networking functions. Human beings are more and more central in technical systems they use and have an active role in their operations along with the Internet devices, through which they communicate, they become the leading actors of a complex socio-technical ecosystem. In this ecosystem, users should not change their lifestyle and adapt themselves to technological requirements but it is the technology that has to take care of people [46]. Personal devices are "key nodes" of the network and make decisions about network functions and data dissemination, by also exploiting human behaviour modeling in physical world. In many cases, network services will be focused on devices, in the sense that they will "take the initiative" and determine how the network must be configured and managed to meet their requirements. They make it possible to dynamically configure the network to access nearby devices and to decide how to use network resources.

To sum up, the characteristics that underlie the IoP paradigm are:

- IoP is human-centric and the IoP algorithms should be based on quantitative models of individual and social human behaviour.
- IoP is device-centric, users' devices are seen as "central IoP nodes", which are proxies of humans in the cyber world.
- IoP is data and computing-oriented, IoP will naturally include the primitives that deal with data management and data computation.
- IoP is self-organizing, IoP users' devices can establish spontaneous and infrastructure-free networks with nearby devices.
  Because of its characteristics, the most profitable paradigm of self-organizing network for IoP is the opportunistic network [16, 18]. In an opportunistic network like this, nodes are personal devices and the network is focused on people and their

devices. The opportunistic network imitates, in the cyber world, the way information spread in the physical world through contact between people. To achieve this goal, personal devices should contribute part of their resources to the dissemination process (crowd sensing, sensing people centric). To develop effective algorithms for data collection in the cyber-physical world, cognitive heuristics have been proposed [28]. These heuristics are effective in the dissemination of data in the cyber-physical world where each node has only a partial knowledge of its environment.

2.2 *Mobile Edge Computing*

Due to the advancement in technologies, mobile sensing and wireless communication, the introduction and development of innovative applications have been encouraged in the last years [1, 30]. Despite the computing and storage capabilities of smart devices have increased over recent years, they are still limited in computational and battery resources in comparison with requirements which are increasingly demanding [25]. Cloud represents the current paradigm for the delivery of services to a massive number of users and it is based on the concentration of huge computation and storage resources in centralized datacentres. However, this way of offloading computation task often leads to unacceptable delay and heavy backhaul usage [4, 5, 29, 66]. MEC is a promising paradigm able to provide a capillary distribution of cloud computing capabilities to mobile devices in a decentralized manner and it is identified as one of the key pillars of 5G [25, 35]. It brings application hosting from centralized data centres to the network edge, closer to users and the data generated by applications [29, 65]. It makes it possible the execution of delay-sensitive and context-aware applications in close proximity of end users while alleviating backhaul utilization and computation at the core network. The main characteristics of MEC are the following [65]:

- Proximity: MEC servers are deployed in close proximity of end devices.
- Low latency: moving data offloading from core network. MEC speeds up application response.
- High bandwidth: the transmission of data between MEC servers and smart devices fully exploit the bandwidth of access network gaining speed.
- Location and network context awareness: MEC servers use received information from smart devices within the local access network to determine specific locations.
- Mobility awareness: mobility is a crucial feature of ubiquitous smart devices. When devices move along, tasks may be offloaded to different MEC servers.
- Heterogeneous resource synergy: to deal with a huge amount of task demands, it is necessary to exploit both cloud and edge computing. In addition, it is necessary synergizing heterogeneous resources, such as computing, caching, communication, etc.

- Real time response: the latency of the offloading process is not only affected by the states of various resources, but also by the strategies of other devices, thus designing computation offloading schemes that can guarantee low latency is not trivial.

  The main advantage of a system that employs MEC is the possibility to enable the typical 5G applications, such as vehicular communication, remote control of robots or machinery, interactive gaming, virtual reality, IoT, Internet of Medical Things (IoMT) [34] and healthcare [27, 37] reducing latency availability of both data and services to the end-users and providing high bandwidth and computing agility in the computation offloading process [2, 25, 66].

  Despite the benefits deriving from the decentralization of cloud computing infrastructure, there are new challenges and open issues deriving from this paradigm [61]:
- Resource management: the computing and storage resources in individual MEC platform are limited and may be able to support a constrained number of application with moderate needs of such resources.
- Interoperability: MEC infrastructure owned by different network providers should be able to collaborate with each other.
- Service discovery: exploiting the synergies of distributed resources and various entities requires discovery mechanisms to automatic discovering of the heterogeneous resources and synchronization.
- Mobility support: solutions for a fast process migration may become necessary.
- Fairness: ensuring fair resource sharing and load balancing is also an essential problem.
- Security: MEC servers could be vulnerable to attacks such as Distributed Denial of Service (DDoS) and issues related to users' privacy- protection mechanisms.

  Due to the limited capacity of each server, edge components may need to compete for the task offloading. In this way in [48] it is explored the positive synergy among MEC and Game Theory (GT) in use cases of incomplete information including aspects of learning, cooperation and social connections in a multi-tiered infrastructure where each layer represents local computing, edge and remote servers. In [61] it is demonstrated that collaborative MECs give better results in term of execution time, latency and faster results. In [65] authors provide a game theoretic-approach for the computational offloading based on incentives that reduces smart devices energy consumption and task execution time. A comparison, through a game-theoretical multiuser computation offloading, between local computation and offloading, is explored in [55] where it is demonstrated that the proposed method achieves better results compared to local computation. A game theory-based power control approach for improving performances, is proposed in [42]. In [31] a Bayesian game theoretic approach is analysed as a guideline for the design of computational offloading strategies. In this paper we consider MEC nodes in a multiplex networks able to cooperate to dynamically distribute tasks, in line with the behavioral dynamics of people in the social multiplex network.

2.3 Multiplex Networks

Over the last decade, as a result of one of the major efforts of the multidisciplinary scientific research, ranging from physics to engineering, the most interesting investigated aspect is that there is a huge number of natural and man-made systems that can be structured as networks [8] with constituents considered as nodes, and interactions modeled as links [9]. A multitude of properties have been deeply explored and they have become the keywords of the most intriguing research studies involving several disciplines and application fields [38–40], starting from the bio-inspired approach which made it possible to include modeling features such as self-organization, self-protection and adaptive dynamics, with the aim to design networks as complex systems [57, 60].

It is well known that a network is a clear representation of entities along with their interactions and global connectivity. Network theory inspects the topology and the structural patterns of interactions among elements of a network. It can be applied to large communication systems, transportation infrastructure [45, 62], biological systems and brain networks [14] and a variety of social interactions structure and phenomena, such as spreading, cooperative and contagion dynamics [8, 23, 58]. Society, telecommunication networks and the human brain represent examples of complex multi-relational systems, whose behavioral dynamics emerge from the analysis of their representative non-trivial structural organizations, non random patterns and interactions among their constituents [13, 45]. In reference to the aforementioned networks, they exhibit specific complex topological features, showing, for example, small-world properties [63]. It implies an average topological distance between nodes increasing logarithmically or slower with the number of nodes, or a power-law behaviour, which characterises the presence of scale-free degree distribution and a statistical abundance of “hubs” with a large number of connections k compared with the average degree value k [6]. These macroscopic properties put in evidence the topological complexity of the systems which many real world networks display in addition to other features, such as heterogeneity of nodes, intensity and types of interactions. To shed light macroscopic, microscopic and mesoscopic points of view, and to obtain the encoding of complexity and other statistical resolution parameters and analysis, the multilayer networks represent the suitable network mathematical formalism. The elementary unit, charted into a network node and the interaction of pairs of nodes is traditionally represented as a connection with a weight. In many cases the representation in which all links are treated on such an equivalent way can be a constraint and it results in loosing details and information, leading to an incomplete description of phenomena that characterize real-world networks.

Some issues related to social networks structure or economics financial networks, viral marketing, gossiping, epidemic spreading and contagion, have been not properly understood, since the traditional formalism assumed that the interactions or relationships take place at the same level. Yet, in reality pairs of interacting elements are connected through multiple types of links, with a multi-scale resolution in space and time. Multilayer networks have been able to generalize the traditional network theory developing a novel framework to study graphs where elements are nodes with several different layers of connections, to be taken into account to properly describe the connectivity and dynamic processes on the system. Multilayer networks incorporate multiple levels of interactions in which each entity, represented as physical node has different manifestations in specific layers, indicated as "state node". Each layer describes different categories of connections, thus considering the multiple relationships between nodes that can be different for relevance, context and meaning [9, 13]. Such a change of paradigm leads to take into account more knowledge than the aggregate of a series of single layers, including links into different groups, the nature of relationships and nodes which belong to each layer, although in some cases it is possible to debate on the possible compressibility [21]. A multilayer network is represented as a family of graphs, as defined in [13]. Graphs can be weighted or unweighted, directed or undirected, and are characterized by a set of nodes and edges. In each layer we have a different set of interconnections between nodes, called intra-layer connections, in contrast with the interconnection among layers, called inter-layer interactions. The concept of multilayer extends other mathematical objects defined as particular type of multilayer networks, such as multiplex networks, networks of networks, interdependent networks, hypergraphs, and many other. Specifically, multiplex networks describe a large number of networks, from social to biological, where a set of node is connected by links of different nature in different layers and the only possible type of inter-layer interactions is that with its counterpart nodes in the rest of layers [9, 13]. The multiplexity represents an extra dimension of analysis which allow us to unveil dynamics and non-trivial phenomena, through multiple channels of connectivity, and providing the more natural description for systems in which nodes have a different set of neighbours in each layer [23]. In the multiplex framework, taking into account the inter-layer interactions between the same nodes at different layers has deep dynamical consequences and gives rise to unexpected emergent phenomena [23]. Each type of interaction can be characterized by a given cost, distance or weight, proving that treating all the links as equivalent in an aggregate structure, results into losing knowledge [10, 45]. The multiplex structure and the referred analysis able us to quantify the information encoded in structure, dynamics and connectivity of the complex system. To show the hidden behavior of complex systems we need to investigate of a multi-dimensional network representation, cause it leads to unveiling interesting structural properties and interdependence, to understand emerging phenomena such as cascading failures, super-diffusion, spreading and epidemic dynamics and evolutionary game-theoretic strategies, to learn and predict collective dynamics in order to design control mechanism and cognitive environment. Thanks to the introduction of multiplex dimension, we acquire a deeper level of analysis in structural terms that can be achieved also by exploring connectivity through weighted multiplex networks so that links between the nodes not only are distinguished by the kind of interaction linking the nodes, but also by the cost or intensity reflecting the importance. Having such a mathematical representation of complex structure, in addition to mathematical algorithms of complex networks, makes us able to extract information and knowledge from systems, to understand cognitively phenomena, identify mechanisms and inspire design of systems, able to dynamically learn to evolve.

2.4 Evolutionary Game Theory (EGT)

The GT is a branch of applied mathematics that analyses conflict situations and searches for competitive or cooperative solutions using mathematical models. It is designed to address situations in which the outcome of individuals' decision depends not just on how they choose among several options, but also on the choices made by the people they are interacting with [26]. EGT originally was born to apply models belonging to the traditional GT to the study of genetics, adaptation and frequency with which a gene appears in a population [23]. EGT differs from classical theories because it focuses more on the dynamics of changing strategies than on achieving an equilibrium. The game is not only a model of interactions among individuals, but also a tool to investigate which behaviours emerge and are able to persist within a population [23]. In particular, EGT has been used extensively to study the problem of cooperation [50]. The purpose of these theories is to use mathematical models to describe and predict what will happen in a game as in a real situation. These studies essentially have a dual role: firstly, a positive role, which consists of interpreting reality and trying to explain why, in specific context, an individual acts in one way rather than in another; the second role, prescriptive, that aims to determine which outcome will arise from the encounter between two or more individuals [49]. At the base of GT and EGT there is obviously the definition of game.

A game is a model of interaction between decision makers where each one plans his actions simultaneously. In each game there are [26]:

- A set of N participants, called players;
- Each player has at his disposal a set $A_i = a_{i1}, a_{i2}, \ldots, a_{in}$ of any possible actions or strategies;
- Together with each strategy, each player receives a payoff that numerically characterizes the player's preference.

In the analysis of the evolution of cooperation it is useful exploit social dilemmas [33], which are two-players and two-strategies games. Social dilemmas are situations in which collective interests are at odds with private interest. They describe situations where the fully selfish and rational behaviour leads to an outcome smaller than the one the individuals would obtain if they acted collectively [15]. Social dilemmas represent the tension between the benefit of the individual and the common good. The two possible strategies are cooperation (C) or defection (D), where cooperating means contributing to the benefit of the whole population paying a cost and defection means being selfish and not paying any cost and relying on the cooperation of other player [23]. To sum up, through the evolutionary game theory, we show the role played by the cooperation of MEC nodes in the multiplex network to provide a multi-services environment with a low blocking probability in response to task offloading problem.

3 Model

3.1 Scenario

The assumed scenario is an innovative smart AAL that provides a multi-service environment towards patients, people with recognised frailty syndromes [51], leveraging the advantages related to 5G technology, mobile edge computing, and complex networks algorithms. Data can be transmitted between devices (such as medical devices, IoT, hand-held devices, etc.) in short range through D2D communication [32]. In this assumed scenario, from one hand, patients are represented as social network nodes linked to each other through different ties (such as real and virtual). Moreover, from this network we can extract big data related on information diffusion, collective awareness, behavioral dynamics and service requests. From the other hand, we have MEC, which guarantees the providing of service in the multi-service environment.

Considering that in AAL we can have different types of service requests and that the 5G network is expected to provide differently eMBB (enhanced mobile broadband) and uRLLC (ultra reliable or low-latency communications), we represent the MEC as nodes of a multiplex network, layered as service-based. In this multiplex network formalization we analyze the fitness of statistical parameters which rules the cooperation of MEC nodes, to guarantee the service requirements.

3.2 Social Multiplex Modeling for AAL and MEC Networks

In this section, we describe the first step of our modeling procedure which consists of defining a social weighted multiplex network [22]. We highlight the importance of including multiple relationships between users in a multiplex network. In the first multiplex, each layer corresponds to a different type of social interaction between users in the AAL (real and virtual layers). In the second multiplex, each layer corresponds to a different type of interaction between MEC nodes, based on service type, eMBB or uRLLC. Thus, let us consider two multiplex network, M_1 and M_2, of M layers $\alpha = \{1, \ldots, M\}$ and N nodes $i = \{1, \ldots, N\}$. A multiplex network is a set of M weighted networks (or layers) $G_\alpha = (V, E_\alpha)$. The set of nodes V is the same for each layer, whereas the set of links E changes according to the layer [45]. Each network G_α is described by the adjacency matrix, denoted by A^α with elements a^α_{ij}, where $a^\alpha_{ij} = w^\alpha_{ij} > 0$, if there is a link between i and j with a weight w_{ij}, otherwise $a^\alpha_{ij} = 0$. In the multiplex M_1, we include weights as function of the discrepancy of patients awareness on its own health conditions and the assortativity with other patients [58, 64]:

$$(w^\alpha_{ij})^{M_1} = h_{ij}|\Delta a_{ij}| + 1 \tag{1}$$

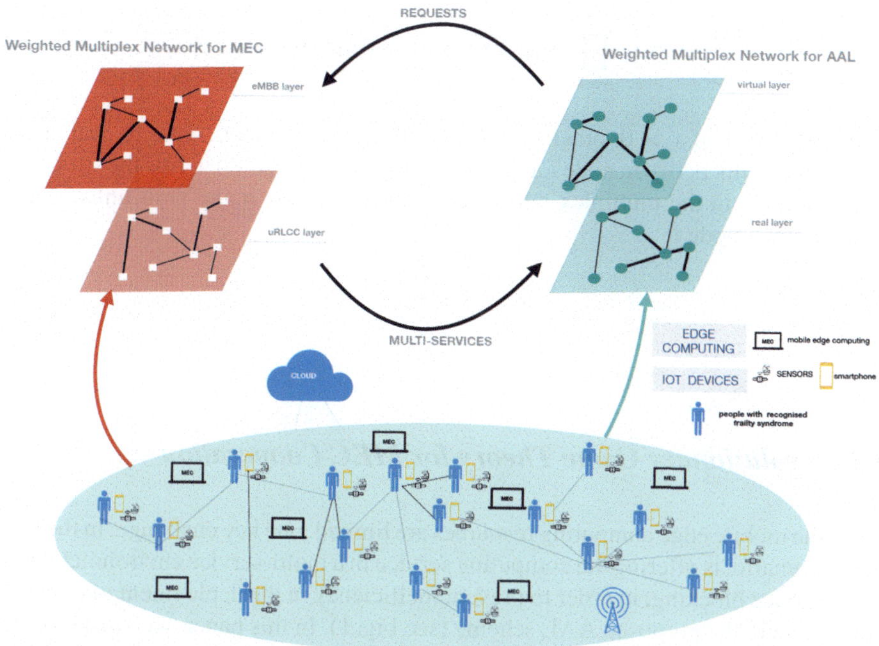

Fig. 1 Multiplex network representation of mobile edge computing and ambient assisted living. The figure illustrates the networks representation in two multiplex networks, M_1 and M_2. Starting from the scenario of a Smart Ambient Assisted Living which includes mobile edge computing, internet of things devices, patients, that are people with frailty conditions with hand-held devices. The two multiplex are interdependent forwarding from one side the services requests and from the other side the multi-services environment

In the multiplex M_2, weights are function of p, that represents the maximum processing capacity that a node assigns to an application of the specific layer, based on requirements of uRLLC and eMBB services [36]:

$$(w_{ij}^{\alpha})^{M_2} = 1 + |p_i^{\alpha} - p_j^{\alpha}| \tag{2}$$

Since in a multiplex network a pair of nodes can be connected in multiple ways, the concept of multilink can be defined as a mesoscale property, which represents the pattern of connections between two nodes [47]. In fact, networks are also characterised by non-trivial structural patterns not only at the level of single-node but also at the level of sub-graph, which adds level of richness to deepen inside structural properties [9]. Let us consider in the M_2 the vector $\overrightarrow{m} = (m_1, m_2, \ldots, m_{\alpha}, \ldots, m_{M_1})$, defined as a multilink, in which every element m_{α} can have only two values $m_{\alpha} = 0, 1$. As showed in [47], we can introduce the multiadjacency matrices $A^{\overrightarrow{m}}$ with elements ${A_{ij}}^{\overrightarrow{m}}$, and the multidegree $k_i^{\overrightarrow{m}}$, given by the following equations:

$$A_{ij}^{\vec{m}} = \prod_{\alpha=1}^{M} [\theta(a_{ij}^{\alpha})m_{\alpha} + (1 - \theta(a_{ij}^{\alpha}))(1 - m_{\alpha})] \tag{3}$$

where $\theta(x) = 1$, if $x > 0$, otherwise $\theta(x) = 0$. These elements serve to evaluate if exists a multilink between node i and node j. To estimate in these terms the structural role of a node in the multiplex network, we evaluate how many multilinks $\vec{m}$ are incident upon node i:

$$k_i^{\vec{m}} = \sum_{j=1}^{N} A_{ij}^{\vec{m}} \tag{4}$$

3.3 *Evolutionary Game Theory for MEC Cooperation*

Since the mobile edge computing resources are limited, one key challenge, in the proposed scenario, is offering the computing service in a multi-service environment with a low service blocking, in order to prevent inefficiency in the deployment of computing system in the proposed AAL scheme (see Fig. 1). In this paper, we introduce the evolutionary game-theoretic approach to deal with this issue by enabling cooperation between MEC nodes in the multiplex network. We focus on exploring the evolution of cooperation on the multiplex network M_2 of MEC nodes, in order to reduce the blocking probability [11]. We analyze different social dilemmas, the iterated forms of Prisoner's Dilemma game (PD), the Snowdrift game (SD), the Stag-Hunt game (SH) and the Harmony game (HG). Since these dilemmas are two-strategies games, with different features, as specified in the payoff matrix, MEC nodes can choose between two strategies: cooperate (C) or defect (D) (Table 1):

Where R represents the reward gained by a co-operator playing against another co-operator, S is the sucker payoff obtained by a co-operator that plays against a defector, T is the temptation payoff obtained by a defector when he plays against a co-operator, and P represents the punishment for the mutual defection. The selection of the matrix parameters enables the definition of several games according to their evolutionary stability [54].

If R > S and R > T > P the game is the HG and the final state of a population that plays this game is the total cooperation. The opposite situation, with T > R > P > S, is represented by the PD; T > R > S > P yields the SD, finally R > T > P

Table 1 Payoff matrix

	C (Cooperation)	D (Defection)
C (Cooperation)	R	S
D (Defection)	T	P

> S corresponds to the SH [44]. Cooperating means that two MEC nodes exchange each other generic computation requests, linked to a specific application, when one of them is temporary overloaded [11]. However, there could be MEC nodes that choose to defect. In fact, if a node i, at the edge of the multiplex network M_2, decides to work on isolation, it can use a buffer, with size k to store computation requests when its CPU is busy. Consequently, it can have the possibility of dropping requests when this buffer is full. However, if multiple MEC nodes can come to rescue of each others, through cooperative behaviors of the node's neighborhood in the multiplex network M_2, dropping probabilities will be reduced, exploiting both multiplexity and EGT. In order to evaluate the performance of our proposed approach, we consider a metric: P_i, namely the blocking probability of the service [11], i.e., the probability that a service request is not executed by either the MEC node (where the request arrived) or by the cooperative node on the weighted multiplex network M_2. Thus, it is given by:

$$P_i = \frac{1}{k_i^{\vec{m}}} \sum_{j=1}^{N_2} \sum_{R_j=T_2}^{R_k} \pi_{R_k R_j} \cdot \pi_{R_j^*} \tag{5}$$

where, $k_i^{\vec{m}}$ is the multidegree of node i, R_j represents the service requests number accepted by node i, of the total service requests sent from weighted multiplex M_1. R_k is the service requests number equal to k, buffer size, which is the maximum requests number allowed. We consider a scenario where the two multiplex are interdependent and, for the sake of simplicity, we assume that the flows of requests from weighted multiplex network M_1, arrive according to a Poisson distribution with rate of λ_i. We define the blocking states of a MEC node as a state where the incoming requests cannot be stored or shared in the neighbors. Differently from the case of a single network, where we can consider just only one type of service application, the introduction of multiplex dimension, layered as service-based, enables MEC nodes to gain from cooperation in different services and in the different layers. This means that, for example, a node that has a high blocking probability for one service benefits from the cooperation of his neighbors, getting the opportunity to provide also other services. The multiplex dimension enables us to evaluate the global blocking probability in multi-service scenario. We aim at quantifying the gain of leveraging the multiplex representation of both AAL and MEC network nodes, the collective behaviors of the multiplex M_1 which produce requests for the MEC network and the evolution of cooperation in multiplex network M_2, to efficiently provide a multi-service environment. We assume, in this paper, that the collective dynamics of the weighted multiplex network M_1 constitute an input to the other weighted M_2 in terms of service requests R_j, which impact the blocking probability. The final aim is to minimize the blocking probability in this given scenario leveraging the evolution of cooperation of the MEC nodes.

4 Simulation and Results

Simulations have been conducted taking into account the two weighted multiplex networks M_1 and of M_2 both consisting of $M = 2$ layers, and following the analytical model described in Sect. 3. The M_1 is referred to the multiplex representation of patients networks in AAL with $N_1 = 1000$ nodes, and the M_2 is referred to the multiplex representation of MEC service-based network, with $N_2 = 100$ nodes. Each layer of both weighted multiplex networks is modeled by considering the Scale- free network as network topology [7]. The reason behind this choice is due, from one hand, to its ability to give a boost to spreading collective dynamics [58] and, from the other hand, it represents the most suitable network topology for the emergence of cooperation [22, 24]. In this way, we are able to investigate how such synthetic network, characterized by controlled topological properties, allow us to derive the potential gain both in terms of behavioral dynamics of the patients multiplex network and in terms of MEC cooperation to provide uRLLC and eMBB services. In Fig. 2 we display the M_1 multiplex representation of patients networks in AAL. Notably, the choice of SF topology is induced by that SF is inherently heterogenoeus, strictly resembling real-world networks displaying a skewed statistical distribution deriving from the preferential attachment rule ("*rich get richer*") [53]. In Fig. 3 we show the aggregate network of the multiplex representation of MEC service-based network, underlining the multidegree, based on the analytical model described in Sect. 3, which

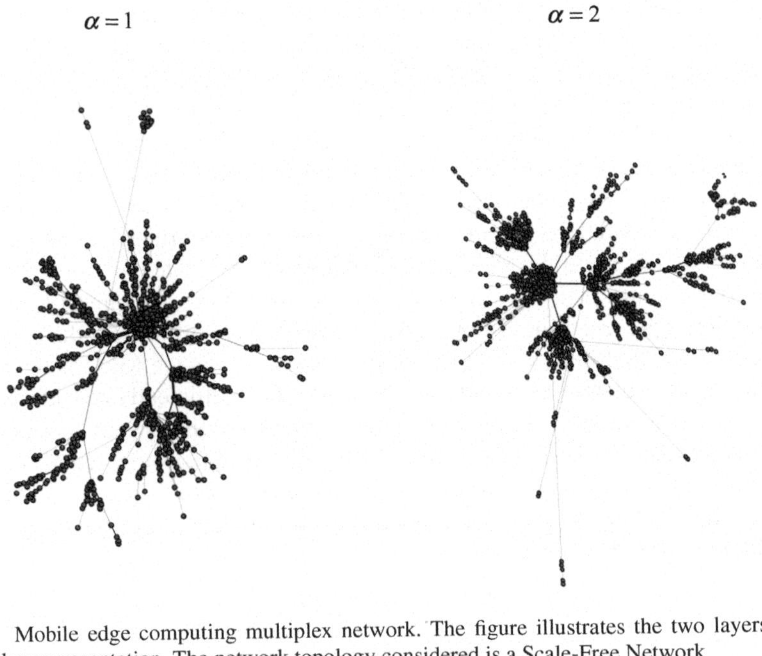

Fig. 2 Mobile edge computing multiplex network. The figure illustrates the two layers of the multiplex representation. The network topology considered is a Scale-Free Network

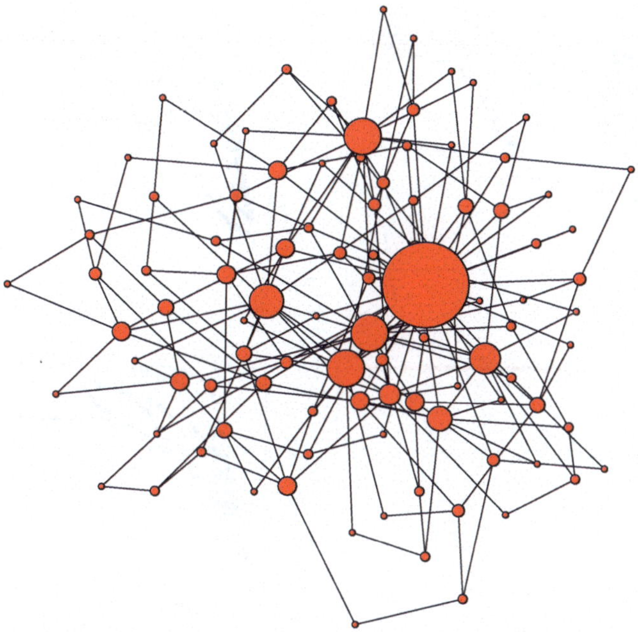

Fig. 3 Mobile edge computing aggregate network. To evaluate the multilink for each node of the multiplex network it is convenient to construct the aggregate network $\hat{G}$, formed by the N_2 nodes of MEC multiplex network. The size of nodes represent the measure of the multidegree

indicates how many multilinks are incident upon a node i. In Fig. 4 we show the density plots to highlight the evolution of cooperation based on four different social dilemmas, respectively PD, SD, HG, SH in (a)-(b)-(c)-(d). We have simulated the evolutionary dynamics in all the different configurations of dilemmas for a number of rounds such that a dynamical steady-state was reached. Results confirm that SF is the most suitable network topology for the emergence and maintenance of the cooperation. Strikingly, we show that the HG, and secondly SD, achieve faster the highest density of cooperators than in the other cases, PD and SH, as expected from literature [44]. In Fig. 5 we shed light on the impact of different social dilemmas on the blocking probability of MEC nodes in the M_2 weighted multiplex network. A node having a high blocking probability for one service gains in capacity from the cooperative behavior of his MEC nodes neighbours in multiplex to provide computing for the global multiservices environment. We quantify the role of different social dilemmas, PD - SD - HG - SH (see respectively (a)-(b)-(c)-(d) of Fig. 5) on the global blocking probability in multiservices scenario of MEC nodes, depending on their multidegree and the number of cooperators in the multiplex. We find out that the evolution of cooperation in the weighted multiplex network of MEC nodes influence the probability that a service request (uRLLC or eMBB) is not executed by either the node or the cooperative neighbouring one, and it decreases as the number of cooperators increases. The multidegree of MEC nodes, displayed in Fig. 3 and defined

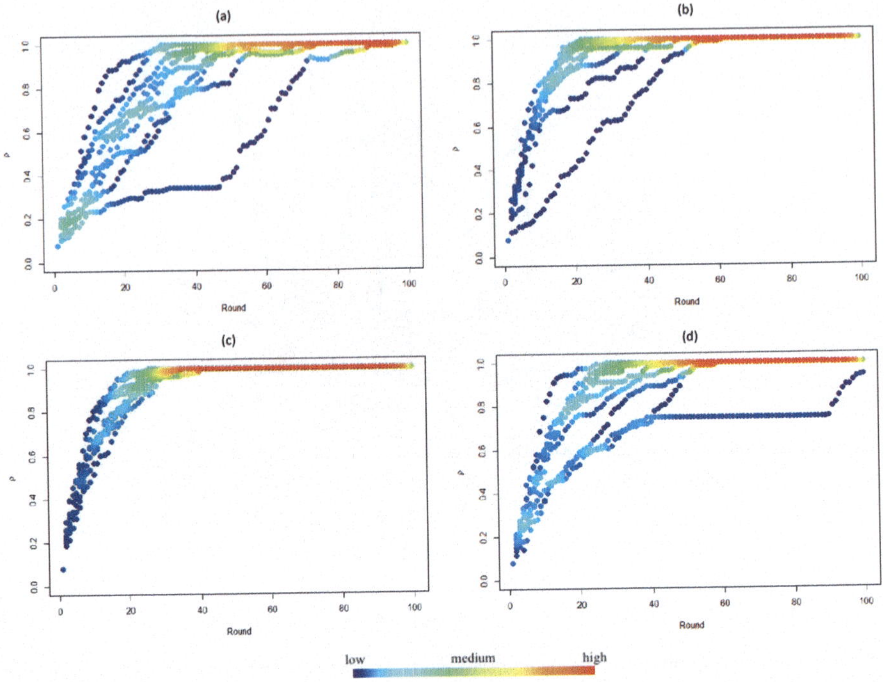

Fig. 4 Density plots. The figures illustrate the density of cooperation in various social dilemma, against the rounds of the game: we show the density ρ, ranging from blue (lowest) to red (highest) in the following games: **a** PD - **b** SD - **c** HG - **d** SH, with $M = 2$ layers in the weighted multiplex network M_2

in Sect. 3, represents the potential impact of service load on the MEC nodes, but also the potential neighborhood with whom to cooperate in order to provide different services. Thus, it is able to trigger a cooperative mechanism. If a MEC node, in the weighted multiplex network M_2, decides to work on isolation, which means that it maintains its interactions but without cooperative approach, potentially it could reduce its blocking probability by increasing the buffer size with a cost [11]. When there is not variation of buffer size, for that node, the higher is its multidegree the higher is its blocking probability, since in this case its neighbourhood, represented by the number of incident multilinks (see Sect. 3), is large and selfish thus it can increase its temporary load peaks. We find out, that the gain of the evolution of cooperation is more striking for the nodes with high multidegree measure as showed in Fig. 5. The evolution of cooperation in weighted multiplex networks, with scale-free structure as network topology of layers, is able to uniform the blocking probability regardless of the number of incident multilinks in a node, when dynamical steady-state is reached.

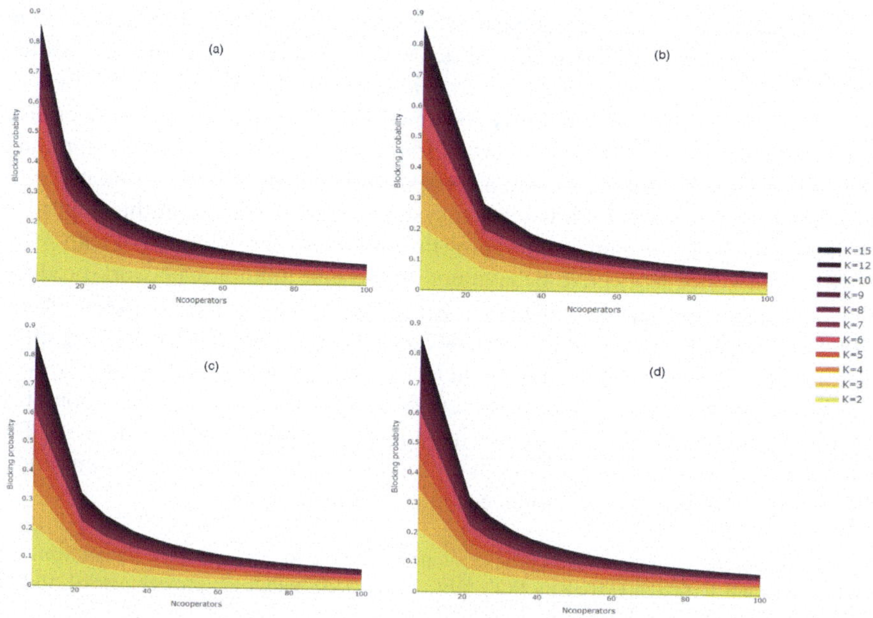

Fig. 5 The impact of evolutionary game theory on blocking probability of MEC nodes. Plots display the blocking probability of nodes against the number of cooperators in the different social dilemmas: **a** PD - **b** SD - **c** HG - **d** SH. We highlights the blocking probability of nodes with various measures of multidegree, identified by the colors ranging from yellow (the minimum value) to black (the maximum value)

5 Conclusion and Future Works

Our idea of including the multiplex representation and the EGT, which enable cooperation of MEC and meets up needs of an AAL of frail people, represents an approach to evaluate the impact of evolutionary dynamics and complex networks in a healthcare scenario. The proposed framework and model encompasses, from one hand, the collective dynamics of the social network representation of patients and, on the other hand, the cooperation of MEC in multiplex network, which provides services to AAL. People can share contents improving knowledge on their own conditions within the AAL. In terms of complex networks this is crucial to improve collective awareness, giving a non-trivial impact on the treatments and the life conditions of people [64]. For example, in the case of frailty associated to life-changing illness [12], the emotional and mental impact is vast and can be long-term [43, 58]. Although any illness can trigger depression, the risk of chronic illnesses and depression can cause possible aggravation in daily life conditions. Thus, patients with life-changing illnesses which share their experience through interactions in a real and virtual network, represented as a social multiplex network, can improve their outcomes [64]. The multiplex network of patients, if properly monitored, such as in a smart cognitive AAL, allows them to share clinical experience and condition, producing, at the same time, data about collective behaviors, needs, service requirements and habits. Through the same social network, this can enable a positively sharing of experiences rather than driving a possible distress contagion [58]. To have a dynamic cognitive multi-service environment for a network of frail people, we need to move computing at network edge, closer to users, and a cooperative behavior is needed among computing nodes, to guarantee multi-service system. For this reason, in this paper we aimed at quantifying the gain of leveraging the multiplex representation of both AAL and MEC networks. Collective behaviours on the first multiplex network produce services requests for the MEC network, while the evolution of cooperation on the second multiplex network is able to provide an efficient multi-service and cognitive environment. MEC nodes benefit from their own cooperation impacting on blocking probability gaining capacity to provide services. We envision to consider a complex scenario including collective dynamics in patients networks, such as awareness diffusion and mesoscopic aspects, in both multiplex networks, to detect communities, with the aim of improving performances in providing services.

Acknowledgements This work was partially supported by the Research Grants: Italian Ministry of University and Research (MIUR) - PON REC 2014–2020 within the project ARS01_01116 "TALIsMAN".

References

1. Aazam, M., Huh, E.N.: Fog computing and smart gateway based communication for cloud of things. In: 2014 International Conference on Future Internet of Things and Cloud, pp. 464–470. IEEE (2014)
2. Aazam, M., Zeadally, S., Harras, K.A.: Offloading in fog computing for iot: review, enabling technologies, and research opportunities. Future Gener. Comput. Syst. **87**, 278–289 (2018)
3. Ahad, A., Tahir, M., Yau, K.L.A.: 5g-based smart healthcare network: architecture, taxonomy, challenges and future research directions. IEEE Access **7**, 100747–100762 (2019)
4. Alam, M.G.R., Hassan, M.M., Uddin, M.Z., Almogren, A., Fortino, G.: Autonomic computation offloading in mobile edge for iot applications. Futur. Gener. Comput. Syst. **90**, 149–157 (2019)
5. Alfakih, T., Hassan, M.M., Gumaei, A., Savaglio, C., Fortino, G.: Task offloading and resource allocation for mobile edge computing by deep reinforcement learning based on sarsa. IEEE Access **8**, 54074–54084 (2020)
6. Barabási, A.L., Albert, R.: Emergence of scaling in random networks. Science **286**(5439), 509–512 (1999)
7. Barabási, A.L., Bonabeau, E.: Scale-free networks. Sci. Am. **288**(5), 60–69 (2003)
8. Barrat, A., Barthelemy, M., Pastor-Satorras, R., Vespignani, A.: The architecture of complex weighted networks. Proc. Natl. Acad. Sci. **101**(11), 3747–3752 (2004)
9. Battiston, F., Nicosia, V., Latora, V.: The new challenges of multiplex networks: measures and models. Eur. Phys. J. Spec. Top. **226**(3), 401–416 (2017)
10. Battiston, F., Perc, M., Latora, V.: Determinants of public cooperation in multiplex networks. In: New Journal of Physics (2017)
11. Beraldi, R., Mtibaa, A., Alnuweiri, H.: Cooperative load balancing scheme for edge computing resources. In: 2017 Second International Conference on Fog and Mobile Edge Computing (FMEC), pp. 94–100. IEEE (2017)

12. Bhatti, Z.U., Salek, M., Finlay, A.Y.: Chronic diseases influence major life changing decisions: a new domain in quality of life research. J. R. Soc. Med. **104**(6), 241–250 (2011)
13. Boccaletti, S., Bianconi, G., Criado, R., Del Genio, C.I., Gómez-Gardeñes, J., Romance, M., Sendiña-Nadal, I., Wang, Z., Zanin, M.: The structure and dynamics of multilayer networks. Phys. Rep. **544**(1), 1–122 (2014)
14. Boccaletti, S., Latora, V., Moreno, Y., Chavez, M., Hwang, D.U.: Complex networks: structure and dynamics. Phys. Rep. **424**(4), 175–308 (2006)
15. Capraro, V.: A model of human cooperation in social dilemmas. PLoS One **8**(8), e72427 (2013)
16. Casadei, R., Fortino, G., Pianini, D., Russo, W., Savaglio, C., Viroli, M.: A development approach for collective opportunistic edge-of-things services. Inf. Sci. **498**, 154–169 (2019)
17. Chen, M., Yang, J., Hao, Y., Mao, S., Hwang, K.: A 5g cognitive system for healthcare. Big Data Cogn. Comput. **1**(1), 2 (2017)
18. Conti, M., Mordacchini, M., Passarella, A.: Design and performance evaluation of data dissemination systems for opportunistic networks based on cognitive heuristics. ACM Trans. Auton. Adapt. Syst. (TAAS) **8**(3), 12 (2013)
19. Conti, M., Passarella, A., Das, S.K.: The internet of people (iop): a new wave in pervasive mobile computing. Pervasive Mob. Comput. **41**, 1–27 (2017)
20. David, K., Berndt, H.: 6g vision and requirements: is there any need for beyond 5g? IEEE Veh. Technol. Mag. **13**(3), 72–80 (2018)
21. De Domenico, M., Nicosia, V., Arenas, A., Latora, V.: Structural reducibility of multilayer networks. Nat. Commun. **6**, 6864 (2015)
22. Di Stefano, A., Scatà, M., La Corte, A., Das, S.K., Liò, P.: Improving qoe in multi-layer social sensing: a cognitive architecture and game theoretic model. In: Proceedings of the Fourth International Workshop on Social Sensing, pp. 18–23. ACM (2019)
23. Di Stefano, A., Scatà, M., La Corte, A., Liò, P., Catania, E., Guardo, E., Pagano, S.: Quantifying the role of homophily in human cooperation using multiplex evolutionary game theory. PLOS ONE **10**(10), e0140646 (2015). https://doi.org/10.1371/journal.pone.0140646
24. Di Stefano, A., Scatà, M., Vijayakumar, S., Angione, C., La Corte, A., Liò, P.: Social dynamics modeling of chrono-nutrition. PLoS Comput. Biol. **15**(1), e1006714 (2019)
25. Dinh, T.Q., La, Q.D., Quek, T.Q., Shin, H.: Learning for computation offloading in mobile edge computing. IEEE Trans. Commun. **66**(12), 6353–6367 (2018)
26. Easley, D., Kleinberg, J., et al.: Networks, crowds, and markets, vol. 8. Cambridge university press, Cambridge (2010)
27. Elmisery, A.M., Rho, S., Botvich, D.: A fog based middleware for automated compliance with oecd privacy principles in internet of healthcare things. IEEE Access **4**, 8418–8441 (2016)
28. Gigerenzer, G., Goldstein, D.G.: Reasoning the fast and frugal way: models of bounded rationality. Psychol. Rev. **103**(4), 650 (1996)
29. Granelli, F.: The role of cloud and mec in 5g
30. Guardo, E., Di Stefano, A., La Corte, A., Sapienza, M., Scatà, M.: A fog computing-based iot framework for precision agriculture. J. Internet Technol. **19**(5), 1401–1411 (2018)
31. Guglielmi, A.V., Levorato, M., Badia, L.: A bayesian game theoretic approach to task offloading in edge and cloud computing. In: 2018 IEEE International Conference on Communications Workshops (ICC Workshops), pp. 1–6. IEEE (2018)
32. Gupta, A., Jha, R.K.: A survey of 5g network: architecture and emerging technologies. IEEE Access **3**, 1206–1232 (2015)
33. Iyer, S., Killingback, T.: Evolution of cooperation in social dilemmas on complex networks. PLoS Comput. Biol. **12**(2), e1004779 (2016)
34. Joyia, G.J., Liaqat, R.M., Farooq, A., Rehman, S.: Internet of medical things (iomt): applications, benefits and future challenges in healthcare domain. J. Commun. **12**(4), 240–7 (2017)
35. Kekki, S., Featherstone, W., Fang, Y., Kuure, P., Li, A., Ranjan, A., Purkayastha, D., Jiangping, F., Frydman, D., Verin, G., et al.: Mec in 5g networks. ETSI White Pap. **28**, 1–28 (2018)
36. Kherraf, N., Alameddine, H.A., Sharafeddine, S., Assi, C., Ghrayeb, A.: Optimized provisioning of edge computing resources with heterogeneous workload in iot networks. In: IEEE Transactions on Network and Service Management (2019)

37. Kraemer, F.A., Braten, A.E., Tamkittikhun, N., Palma, D.: Fog computing in healthcare-a review and discussion. IEEE Access **5**, 9206–9222 (2017)
38. La Corte, A., Scatà, M.: A process approach to manage the security of the communication systems with risk analysis based on epidemiological model. In: 2010 Fifth International Conference on Systems and Networks Communications, pp. 166–171. IEEE (2010)
39. La Corte, A., Scatá, M.: Convergence, security and quality in the ngn. Int. J. Internet Technol. Secur. Trans. **4**(4), 327–343 (2012)
40. La Corte, A., Scatá, M., Giacchi, E.: A bio-inspired approach for risk analysis of ict systems. In: International Conference on Computational Science and Its Applications, pp. 652–666. Springer (2011)
41. Li, M.: Internet of people. Concurr. Comput. Pract. Exp. **29**(3) (2017)
42. Li, N., Martinez-Ortega, J.F., Diaz, V.H.: Distributed power control for interference-aware multi-user mobile edge computing: a game theory approach. IEEE Access **6**, 36105–36114 (2018)
43. Lotfaliany, M., Bowe, S.J., Kowal, P., Orellana, L., Berk, M., Mohebbi, M.: Depression and chronic diseases: co-occurrence and communality of risk factors. J. Affect. Disord. **241**, 461–468 (2018)
44. Matamalas, J.T., Poncela-Casasnovas, J., Gómez, S., Arenas, A.: Strategical incoherence regulates cooperation in social dilemmas on multiplex networks. Sci. Rep. **5**, 9519 (2015)
45. Menichetti, G., Remondini, D., Panzarasa, P., Mondragón, R.J., Bianconi, G.: Weighted multiplex networks. PloS One **9**(6), e97857 (2014)
46. Miranda, J., Mäkitalo, N., Garcia-Alonso, J., Berrocal, J., Mikkonen, T., Canal, C., Murillo, J.M.: From the internet of things to the internet of people. IEEE Internet Comput. **19**(2), 40–47 (2015)
47. Mondragon, R.J., Iacovacci, J., Bianconi, G.: Multilink communities of multiplex networks. PloS One **13**(3), e0193821 (2018)
48. Moura, J., Hutchison, D.: Game theory for multi-access edge computing: survey, use cases, and future trends. IEEE Commun. Surv. Tutor. **21**(1), 260–288 (2018)
49. Myerson, R.B.: Game Theory. Harvard University Press (2013)
50. Nowak, M.A., Sigmund, K.: Evolutionary dynamics of biological games. Science **303**(5659), 793–799 (2004)
51. O'Caoimh, R., Galluzzo, L., Rodríguez-Laso, Á., Van der Heyden, J., Ranhoff, A.H., Lamprini-Koula, M., Ciutan, M., Samaniego, L.L., Carcaillon-Bentata, L., Kennelly, S., et al.: Prevalence of frailty at population level in european advantage joint action member states: a systematic review and meta-analysis. Annali dell'Istituto superiore di sanita **54**(3), 226–239 (2018)
52. Pal, D., Funilkul, S., Charoenkitkarn, N., Kanthamanon, P.: Internet-of-things and smart homes for elderly healthcare: an end user perspective. IEEE Access **6**, 10483–10496 (2018)
53. Pastor-Satorras, R., Castellano, C., Van Mieghem, P., Vespignani, A.: Epidemic processes in complex networks. Rev. Mod. Phys. **87**(3), 925 (2015)
54. Perc, M., Szolnoki, A.: Coevolutionary games'a mini review. BioSystems **99**(2), 109–125 (2010)
55. Qin, A., Cai, C., Wang, Q., Ni, Y., Zhu, H.: Game theoretical multi-user computation offloading for mobile-edge cloud computing. In: 2019 IEEE Conference on Multimedia Information Processing and Retrieval (MIPR), pp. 328–332. IEEE (2019)
56. Saad, W., Bennis, M., Chen, M.: A vision of 6g wireless systems: applications, trends, technologies, and open research problems. In: IEEE network (2019)
57. Scatà, M., Di Stefano, A., Giacchi, E., La Corte, A., Liò, P.: The bio-inspired and social evolution of node and data in a multilayer network. In: 2014 5th International Conference on Data Communication Networking (DCNET), pp. 1–6. IEEE (2014)
58. Scatà, M., Di Stefano, A., La Corte, A., Liò, P.: Quantifying the propagation of distress and mental disorders in social networks. Sci. Rep. **8**(1), 5005 (2018)
59. Scatà, M., Di Stefano, A., Liò, P., La Corte, A.: The impact of heterogeneity and awareness in modeling epidemic spreading on multiplex networks. Sci. Rep. **6**(37105) (2016). https://doi.org/10.1038/srep37105

60. Scatà, M., La Corte, A.: Security analysis and countermeasures assessment against spit attacks on voip systems. In: 2011 World Congress on Internet Security (WorldCIS-2011), pp. 177–183. IEEE (2011)
61. Tran, T.X., Hajisami, A., Pandey, P., Pompili, D.: Collaborative mobile edge computing in 5g networks: new paradigms, scenarios, and challenges. IEEE Commun. Mag. **55**(4), 54–61 (2017)
62. Varga, I.: Weighted multiplex network of air transportation. Eur. Phys. J. B **89**(6), 139 (2016)
63. Watts, D.J., Strogatz, S.H.: Collective dynamics of "small-world" networks. Nature **393**(6684), 440 (1998)
64. Wicks, P., Massagli, M., Frost, J., Brownstein, C., Okun, S., Vaughan, T., Bradley, R., Heywood, J.: Sharing health data for better outcomes on patientslikeme. J. Med. Internet Res. **12**(2), e19 (2010)
65. Zhang, K., Leng, S., He, Y., Maharjan, S., Zhang, Y.: Mobile edge computing and networking for green and low-latency internet of things. IEEE Commun. Mag. **56**(5), 39–45 (2018)
66. Zhang, K., Mao, Y., Leng, S., Zhao, Q., Li, L., Peng, X., Pan, L., Maharjan, S., Zhang, Y.: Energy-efficient offloading for mobile edge computing in 5g heterogeneous networks. IEEE Access **4**, 5896–5907 (2016)

Correlations Among Game of Thieves and Other Centrality Measures in Complex Networks

Annamaria Ficara, Giacomo Fiumara, Pasquale De Meo, and Antonio Liotta

Abstract Social Network Analysis (SNA) is used to study the exchange of resources among individuals, groups, or organizations. The role of individuals or connections in a network is described by a set of centrality metrics which represent one of the most important results of SNA. Degree, closeness, betweenness and clustering coefficient are the most used centrality measures. Their use is, however, severely hampered by their computation cost. This issue can be overcome by an algorithm called Game of Thieves (GoT). Thanks to this new algorithm, we can compute the importance of all elements in a network (i.e. vertices and edges), compared to the total number of vertices. This calculation is done not in a quadratic time, as when we use the classical methods, but in polylogarithmic time. Starting from this we present our results on the correlation existing between GoT and the most widely used centrality measures. From our experiments emerge that a strong correlation exists, which makes GoT eligible as a centrality measure for large scale complex networks.

1 Introduction

SNA studies groups [14] of individuals and it can find an application in a lot of areas such us organizational studies, psychology, economics, information science and criminology [5, 13]. Social Networks (SNs) like Facebook and Twitter have grown exponentially providing new challenges for the application of SNA methods.

A. Ficara (✉)
University of Palermo, Palermo, Italy
e-mail: aficara@unime.it

G. Fiumara · P. De Meo
University of Messina, Messina, Italy
e-mail: gfiumara@unime.it

P. De Meo
e-mail: pdemeo@unime.it

A. Liotta
Free University of Bozen-Bolzano, Bolzano, Italy
e-mail: antonio.liotta@unibz.it

G. Fortino et al. (eds.), *Data Science and Internet of Things*, Internet of Things,
https://doi.org/10.1007/978-3-030-67197-6_3

The definition of the so-called centrality measures represents one of the most important results of SNA. These set of measures describe the role of single individuals (or single connections) with respect to their network of relationships and can be used to identify the most influential people. These people have the potential of controlling the information flow inside a network and, for this reason, they have a great practical relevance. Thanks to the use of the main centrality metrics, such us degree, closeness, betweenness, and clustering coefficient, we can increase our understanding of a network.

When we compute centrality measures on on-line SNs, we are facing the problem about the big size of the data. This problem can be overcome by using a new algorithm called Game of Thieves (GoT) [26]. GoT computes the centrality of both vertices and edges in a network with respect to the total number of vertices. This computation is done in polylogarithmic time, while the classical centrality measures need at least a quadratic time.

GoT owes its name to the protagonists of the game who are a multitude of thieves whose main purpose is to steal diamonds.

The basic idea is to make an overlap between a heterogeneous system like a complex network [30] and a homogeneous artificial system which has two key elements: a group of thieves and a set of vdiamonds (i.e. virtual diamonds). At the beginning, each vertex is artificially endowed with vdiamonds and wandering thieves. If a thief does not carry any vdiamond, his state is "empty". If he carries a vdiamond, his state becomes "loaded". If the thief state is "empty", he wanders in search of vdiamonds. The thief picks randomly a neighbor of the vertex in which it is located, he moves to this new vertex and, if he finds a vdiamond, he fetches it. Then, he follows back the same path used in search of vdiamonds and brings the vdiamond back to his home vertex. At this point the vdiamond becomes available for the other thieves who can steal it. At the beginning of the game, there is the same number of thieves and vdiamonds in each vertex. Then, GoT proceeds in epochs. At each epoch, all thieves move from their current location to the next one. When they find or deposit a new vdiamond, their state ("loaded" or "empty") changes.

Encouraged from the superior performance of GoT respect to the state-of-art algorithms, we decided to investigate whether GoT can be used to compute vertices and edges centrality. This amounts to investigate whether and to which extent exists a correlation between GoT and some classical centrality measures.

Correlation is a bivariate analysis through which we can study the association between two variables. This kind of analysis takes into account the strength of this relationship between pairs of variables and its direction. The value of the correlation coefficient can vary from -1 to $+1$. A perfect degree of association (positive or negative) between two variables is indicated by a value of ± 1. The relationship between pairs of variables becomes weaker when the value of the correlation coefficient goes towards 0. The most used types of correlations are Pearson correlation, Spearman and Kendall rank correlations.

We have done a lot of experiments computing these three correlation metrics on different types of networks both artificial and real. We used three classes of simulated networks: *Erdös Rényi* (ER) random graphs, *small-world* (SW) and *scale-free*

(SF) networks. For each class, we randomly generated different networks which have 1,000–15,000 vertices and 4,970–1,125,545 edges. Then, we have taken into account three networks from real-world: Dolphins (62 vertices, 159 edges, unweighted), High Energy (8,361 vertices, 15,751 edges, weighted, disconnected) and Internet (22,963 vertices, 48,436 edges, unweighted). Our experiments show that there is a strong negative correlation among GoT and the main centrality metrics like degree, betweenness and closeness; while there is no correlation between GoT and the clustering coefficient with the exception of the small-world networks in which we can find a strong positive correlation.

2 Related Literature

Centrality measures describe the position of an individual in a network in relation to the complete network and to the other individuals in the same network. Some centrality metrics identify the most influential and prestigious actors in a network [3, 15, 33, 39], some others indicate the social influence of an individual with respect to others in a network [16], others evaluate the integration of each individual into a network [38]. Most recently, a new algorithm called Game of Thieves has been developed [26]. It is a method which is able to compute the importance of both vertices and edges, which are the elements of a network, and to complete this computation in polylogarithmic time with respect to the total number of vertices.

Degree centrality, closeness centrality, betweenness centrality and clustering coefficient can be considered as the most frequently used centrality metrics. The first three measures were proposed by Freeman [15], whereas the clustering coefficient was defined by Watts and Strogatz [40]. In our work, we try to answer to an often asked, but rarely answered, question that is: are these centrality measures correlated? If there exists a high correlation between the centrality metrics, we can expect they have a similar behavior in statistical analyses and for this reason the development of multiple measures seems to be redundant. If there is not high correlation, we can conclude that they are unique measures which can be associated with different outcomes. But, we are not only interested in the correlation between the most used centrality metrics. We want to answer to an other question: are these centrality measures correlated with Game of Thieves? If we find that they behave similarly, we can use GoT in the computation of individuals' centrality in very large networks, considerably reducing the execution time of this computation.

Many researchers carried out studies on the correlations between centrality measures.

Bolland [2] made a correlation analysis on four centrality measures: degree, closeness, betweenness, and continuing flow. He considered three criteria that are robustness, face validity and sensitivity. He underlined the similarity between closeness, degree and continuing flow and a relative difference between this three indices and the betweenness centrality. The high intercorrelations among the first three indices produced a considerable redundancy for the used dataset which was increased with the introduction of random error into the data. Then, the author chose the continuing flow as the best model and a useful companion to the betweenness.

Rothenberg et al. [32] compared eight centrality measures analyzing people risky behaviors in an area of low prevalence for HIV transmission. These measures were: three forms of information centrality (i.e. measures of centrality which make use of all paths between pairs of points) [36], eccentricity, mean, and median (i.e., three distance measures), and degree and betweenness centrality. Their studies showed an high correlation among these eight centrality measures. In particular, there was an high correlation among the three distance measures and the three information measures, but there was a weaker correlation among these measures and degree and betweenness. Degree and betweenness were highly correlated, but both were less correlated with the three forms of information centrality which were highly correlated among themselves.

Faust [12] used a subset of the data from Galaskiewicz's study [17] regarding relationships between CEOs, clubs and boards and examined correlations among several centrality measures. He used centrality measures such us degree, eigenvector, closeness, betweenness to compute the centrality of an event, and flow betweenness used to identify central clubs. Then, he studied the correlation among these metrics founding correlation coefficients between 0.89 and 0.99.

Valente and Forman [38] discovered two new centrality measures know as integration and radiality. They examined correlations among these two measures, in-degree, out-degree, closeness, betweenness, flow and density. They used the "Sampson Monastery" and the "Medical Innovations" datasets. Their analysis showed that integration was correlated with in-degree and radiality was correlated with out-degree. A further study on the correlation revealed that these new metrics were similar but distinct from closeness, betweenness and flow.

In a more recent study, Valente et al. [37] choose the most commonly used centrality measures such as degree, in-degree, out-degree, betweenness, s-betweenness, closeness-in, closeness-out, s-closeness, integration, radiality and eigenvector. They empirically investigated the correlation among them finding out that degree had the strongest overall correlations. Eigenvector centrality had the next highest average correlation. Similar correlations were founded among betweenness, symmetrized closeness, in-degree and out-degree. The lowest average correlation was discovered between directional closeness measures, in-closeness and out-closeness.

Li et al. [24] first studied the Pearson correlation between centrality measures and the similarity ranking for vertices. Then, they introduced a new centrality measure known as the degree mass. They found that betweenness, closeness, and eigenvector were strongly correlated with the degree, the 1st-order degree mass and the 2nd-order degree mass, respectively, in both artificial and real networks. Then, they demonstrated that eigenvector and the 2nd-order degree mass had a larger Pearson correlation coefficient respect to eigenvector and a lower order degree mass.

Ronqui and Travieso [31] studied the correlation between pairs of centrality measures in two artificial networks and several real networks. Their analysis showed that these metrics were usually correlated. A stronger correlation could be found in the artificial networks with respect to real networks. Moreover, the strength of the correlation between the centrality measures varied from network to network. For this reason, they proposed a centrality correlation profile as a way to characterize networks. This profile consisted of the values of the correlation coefficients between the centrality metrics of interest.

Grando et al. [20] showed through their experiments that vertex centrality measures such as information, eigenvector, subgraph, walk betweenness and betweenness could identify vertices in all kinds of networks with a performance at 95%. Considerably lower results could be achieved using other metrics. In addition, they demonstrated that several pairs of centrality metrics evaluate the vertices in a very similar way (i.e. their correlation coefficient values were above 0.7).

Shao et al. [34] uses degree to approximate closeness, betweenness, and eigenvector. They first demonstrated that rank correlation performed better than the Pearson one in scale-free networks. Then, they studied the correlation between centrality metrics in real networks. At the end, they demonstrated that largest betweenness and closeness vertices could be approximated by the largest degree vertices. This approximation was not valid for the largest eigenvector vertices.

Oldham et al. [29] used 212 different real networks and calculated correlations between 17 different centrality measures. The relationship between these correlations and the variations in network density and global topology was examined together with the possibility for vertices to be clustered into distinct classes according to their centrality profiles. Their analysis showed that there was a positive correlation among the centrality measures. The strength of these correlations could vary across networks, and network modularity played a key role in driving these cross-network variations.

3 Background

3.1 Centrality Measures

Centrality is a core concept for the SNA. A SN is a set of people interconnected by social ties, e.g., friendship or family relationships [19]. It can be represented using a graph $G = (V, E)$ where V is a set of vertices (also called nodes, actors) and $E \subseteq V \times V$ is a set of edges (also called links, ties). A graph is called *undirected* when all the edges are bidirectional, *directed* when the edges have a specific direction. Given a directed edge $e = (u, v) \in E$, we can say that v is the head of e, u is the tail and v is adjacent to u. Specific graph types can be used depending on the specific SN. We can represent a SN like Facebook with an undirected graph because in this case friendship relationships are reciprocal. Instead, we can use directed graphs to describe SNs like Twitter which use following relationships and require the use of edges with a specific direction.

A SN can be also defined as a *weighted graph* $G = (V, E, W)$ where V is the set of vertices, $E \subseteq V \times V$ is the set of edges, and $W : E \rightarrow R_{++}$ is a set of positive weights defined on each edge.

Degree Centrality (DC) [15] is used to evaluate the local importance of a vertex and it is one of the simplest centrality measures; given a vertex u the degree centrality $DC(u)$ of u is as follows:

$$DC(u) = \sum_{w=1}^{v} a_{uw}$$

where v is the number of vertices in G, $a_{uw} = 1$ if and only if there exists $(u, w) \in E$, 0 otherwise.

Betweenness Centrality (BC) [4] measures how important the role of a vertex is in the propagation of informations. Some vertices, in fact, act as bridges between different parts of a graph and for this reason they can block the flow of informations from one region to other. Specifically, the (shortest-path) betweenness $BC(u)$ of a vertex v is the sum of the fraction of all-pairs shortest paths that pass through u and it defined as follows:

$$BC(u) = \sum_{x,y \in V} \frac{\sigma(x, y|u)}{\sigma(x, y)}$$

where $\sigma(x, y)$ is the number of shortest paths between an arbitrary pair of vertices x and y, and $\sigma(x, y|u)$ is the number of shortest paths which connect x and y by passing through the vertex u.

Closeness Centrality (CL) [15] measures the "proximity" between a vertex and all other vertices in a graph G. The closeness centrality of a vertex u is the reciprocal of the sum of the shortest path distances from u to all other vertices in G, normalized by $v - 1$:

$$CL(u) = \frac{v - 1}{\sum_{w=1}^{v-1} d(u, w)}$$

Clustering Coefficient (CC) measures how connected a vertex neighbors are to one another. For unweighted graphs, the clustering of a vertex u, denoted by $CC(u)$, is the fraction of possible triangles through that vertex that exist,

$$CC(u) = \frac{2T(u)}{D(u)(D(u) - 1)},$$

where $T(u)$ is the number of triangles through vertex u and $D(u)$ is the degree of u. $CC(u) = 1$ if every neighbor connected to a vertex u is also connected to every other vertex within the neighborhood. $CC(u) = 0$ if no vertex that is connected to u connects to any other vertex that is connected to u.

3.2 Game of Thieves

Game of Thieves [26] is a new centrality measure to compute the centrality of vertices and edges in a graph $G = (V, E)$, where V is the set of vertices, and E is the set of edges.

As mentioned in Sect. 1, the leading actors in the game are wandering thieves. If a thief carry a vdiamond, his state is "empty". If a thief does not carry a vdiamond, his state is "loaded".

In order to understand how this measure works, we have to define some notation:

- Φ_0^v is the initial number of vdiamonds in vertex $v \in V$ at epoch $T = 0$;
- Φ_T^v indicates the number of vdiamonds in vertex $v \in V$ at epoch T (i.e. after GoT has run for T epochs);
- Ψ_T^e is the number of "loaded" thieves passing through an edge $e \in E$ at epoch T;
- Γ_v is the set of vertices connected by an edge with vertex v, $\forall v \in V$;
- $\Omega_{vu} \geq 0$ is the weight of the edge which connects the vertex $v \in V$ and $u \in V$;
- Y_t is a dynamic list which contains the vertices visited by a thief t, useful to keep the path of t in his search for vdiamonds.

If the state of a thief t is "empty", the following operations will be sequentially performed in any epoch ep:

step 1: a randomly picks a vertex $u \in \Gamma_v$, where v is its actual location, with a probability $p_{vu} = \frac{\Omega_{uv}}{\sum_{v \in \Gamma_v} \Omega_{vu}}$.
step 2: t moves from his home vertex v to vertex u.
step 3: If $u \in Y_t$, then all the vertices after u in Y_t are removed from the list.
step 4: If $u \notin Y_t$, then u is added to the end of Y_t.
step 5: If $\Phi_{ep}^u > 0$, then t takes one vdiamond and changes his state to "loaded".
step 6: Φ_{ep}^u decreases by one vdiamond.

If the state of a thief t is "loaded", the following steps will be sequentially performed in any epoch ep:

step 1: t moves from the last vertex v from Y_t, which is his actual location, to the last but one vertex u from Y_t.
step 2: v is removed from Y_t.
step 3: Ψ_{ep}^e increases by one, i.e, edge e from v to u increases.
step 4: If u is the home vertex of t, t unloads the vdiamond, and sets his state to "empty".
step 5: Φ_{ep}^u increases by one vdiamond.

The game runs for a duration of T epochs. The number of epochs to stop the algorithm is conventionally $T = \log^3|V|$.

Figure 1 shows snapshots of GoT in action on a simple network with 10 vertices. We can observe the thieves' behavior and consequently the number of vdiamonds on each vertex v after $T = \log^3|10| \approx 12$ epochs.

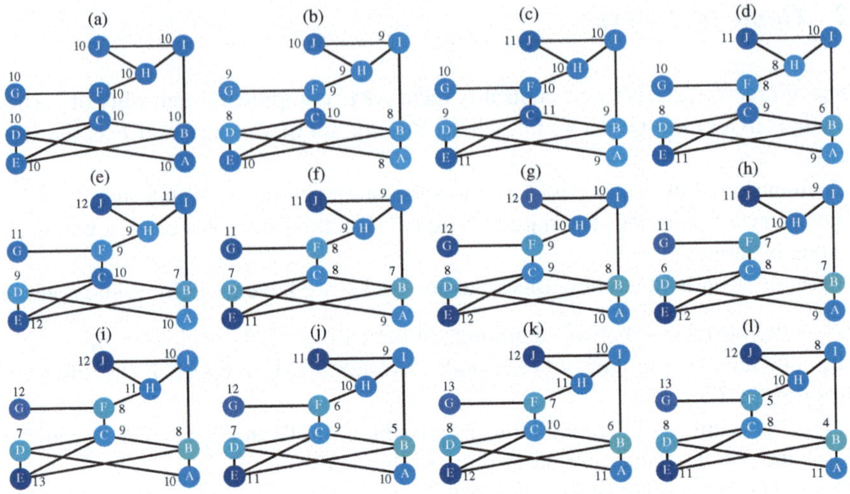

Fig. 1 **GoT in action**. GoT behavior over $T = 12$ epochs on a simple unweighted network with 10 vertices. The initial parameters are $\Phi_0^v = 10$ and 1 thief per vertex. The numbers on the side of each vertex show the number of vdiamonds, Φ_{ep}^v, in any vertex v at epoch ep, where $ep = 1, 2, 3, 4, 5, 6, 7, 8, 9, 10, 11, 12$ epochs in subplots **a, b, c, d, e, f, g, h, i, j, k, l** respectively

When the game stops the centrality of each vertex v is computed as:

$$\bar{\Phi}_T^v = \frac{1}{T}\sum_{ep=0}^{T} \Phi_{ep}^v$$

This measure also refers to the average number of vdiamonds present at a vertex v, after the game has run for a duration of T epochs. An important vertex is indicated by a small Φ_T^v value, while a less important vertex is denoted by a high Φ_T^v value. This is because a lot of thieves visit the most central vertices which will are quickly depleted, while few thieves visit the less central vertices which will not be depleted.

Then, the centrality of each edge e is also computed as:

$$\bar{\Psi}_T^e = \frac{1}{T}\sum_{ep=0}^{T} \Psi_{ep}^e$$

This measure also refers to the average number of thieves who carry a vdiamond (i.e. in "loaded" state) passing through an edge e after T epochs. The most important edges are indicated by a high Ψ_T^e value, while the less important edges are denoted by lower Ψ_T^e values.

The computational complexity of GoT, $O(GoT)$, is bounded by $O(log^2|V|) < O(GoT) < O(log^3|V|)$.

Table 1 Comparison of five centrality algorithms using computational complexity

Algorithm	Computational complexity
Degree centrality	$O(\|V\|)$
Betweenness centrality	$O(\|V\|\|E\|)$
Closeness centrality	$O(\|V^3\|)$
Clustering coefficient	$O(V^2)$
Game of thieves	$O(log^2\|V\|) < O(GoT) < O(log^3\|V\|)$

It's easy to guess from the description of the game that each vertex in the network is independent from the others. An high level of parallelization can be achieved in a traditional parallel computing environment, such as MPI. We can also think about a graph partitioning algorithm in which each vertex or a group of vertices can do their own computations. GoT seems to be a fully distributed algorithm. Table 1 shows how GoT represents a great step forward in terms of time complexity with respect to centrality algorithms such us degree, betweenness, closeness and clustering.

3.3 Correlation Coefficients

The correlation coefficient is a statistical measure of the strength of the relationship between two variables. The values of the coefficient can vary from −1.0 to 1.0. A number greater than 1.0 or less than −1.0 implies an error in the correlation measurement. A correlation of −1.0 means that there is a perfect negative correlation, while a correlation of 1.0 shows a perfect positive correlation. A correlation of 0.0 indicates no relationship between the two variables.

Pearson r correlation coefficient [6] is the most used correlation metric and it measures the degree of association between two linearly related variables. It is computed according to the following formula:

$$r = \frac{s\sum ab - \sum(a)(b)}{\sqrt{[s\sum a^2 - \sum(a^2)][s\sum b^2 - \sum(b^2)]}}$$

where r is the Pearson correlation coefficient, s is the number of observations, $\sum ab$ is the sum of the products of a and b scores, $\sum a$ is the sum of a scores, $\sum b$ is the sum of b scores, $\sum a^2$ is the sum of squared a scores and $\sum b^2$ is the sum of squared b scores.

Spearman rank correlation coefficient [35] is a non-parametric measure of rank correlation. It measures the degree of relationship between two variables. The only hypothesis required is that the two variables can be ordered and, if possible, continued. This coefficient is computed according to the following formula:

$$\rho = 1 - \frac{6\sum d_i^2}{s(s^2-1)}$$

where ρ is the Spearman rank correlation, d_i is the difference between the ranks of corresponding variables and s is the number of observations.

Kendall rank correlation coefficient [23] is a non-parametric test used to measure the strength of association between two variables. If we consider two samples, x and y, where each sample size is s, $s(s-1)/2$ will be the total number of pairings with xy. This coefficient is computed according to the following formula:

$$\tau = \frac{s_c - s_d}{\frac{1}{2}s(s-1)}$$

where s_c is number of concordant pairs and s_d is number of discordant pairs.

3.4 Complex Networks

Random networks. A random network may be described simply by a probability distribution, or by a random process which generates it. The *Erdös Rényi model* is one of two closely related models to generate random networks. There are two variants of the Erdös Rényi model [11]. The first chooses one of all possible networks $G(v, E)$ with v vertices and E edges, where each network has an equal probability. This could be done by choosing V edges from the $\binom{v}{2}$ possible edges. Second variant $G(v, p)$ [18] starts with an initial set of v unconnected vertices and includes edges with probability p. It can easily be deduced that each network with v vertices and E edges is equally likely with probability:

$$p^E(1-p)^{\binom{v}{2}-E}$$

In this paper, we used the second variant $G(v, p)$ of the ER model. In each experiment, we have chosen the number of vertices v between $1,000$ (see Fig. 2) and $15,000$, and a probability for edge creation $p = 0.01$.

Small-world networks. A small-world network [10] is characterized by a high degree of local clustering (like regular lattices). It also possess short vertex-vertex distances. This network model was proposed by Watts and Strogatz [40] and it interpolates between these two extremes by taking a regular lattice and randomly rewiring some of its edges.

Newman and Watts [27] proposed a variation of the Watts and Strogatz model. Given a network defined as a graph $G(V, E)$, where V is the set of vertices and E is the set of edges, the *Newman-Watts-Strogatz small-world model* (NWS) is defined as follows:

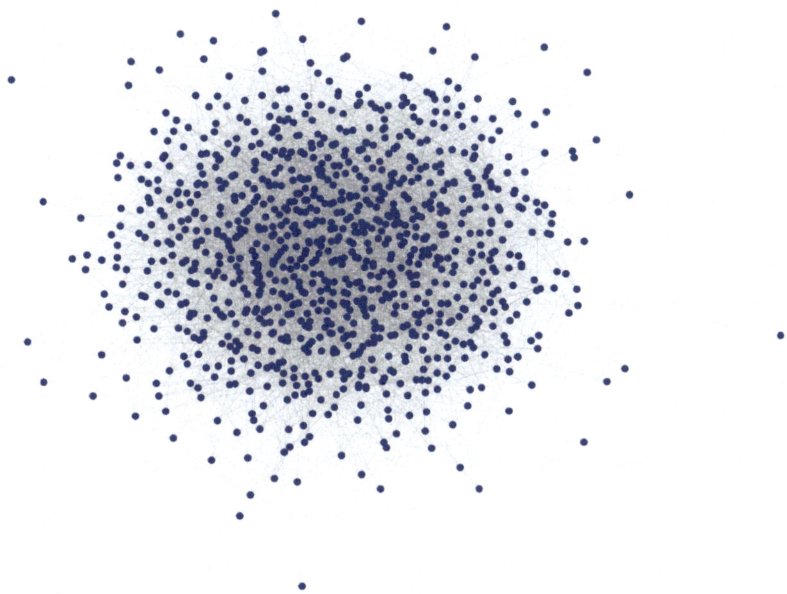

Fig. 2 **Random network.** ER model created using the variant $G(v, p)$ with $v = 1000$ and $p = 0.01$ probability for edge creation

step 1: *Ring Creation.* Creation of a ring over v vertices in which each vertex $u \in V$ is connected with the k closest neighbors. If k is odd, u is connected with the nearest $k - 1$ neighbors.

step 2: *Edge rewiring.* For each edge $(u, w) \in E$, in the underlying v-ring with k nearest neighbors, a new edge (u, w) is added, with randomly-chosen existing vertex w and probability p.

Compared with Watts-Strogatz model, the random rewiring increases the edges number because new edges are added and no edges are removed.

In this paper, we used the NWS model. In each experiment, we have chosen the number of vertices v between $1,000$ (see Fig. 3) and $15,000$, $k = 6$ neighbors with which connect each vertex u in the ring topology, and a probability $p = 0.6$ of rewiring each edge.

Scale-free networks. A scale-free [10] network is characterized by a degree distribution (i.e. the distribution of the number of vertices that have a particular degree) which decays like a power law [1]. Given a network defined as a graph $G(V, E)$, where V is the set of vertices and E is the set of edges, the scale-free network model of *Barabási and Albert* (BA) is defined as follows:

step 1 *Initial condition.* The network consists of v_0 vertices and e_0 edges.

step 2 *Growth.* One vertex u with e edges is added at each step. Time t is the number of steps.

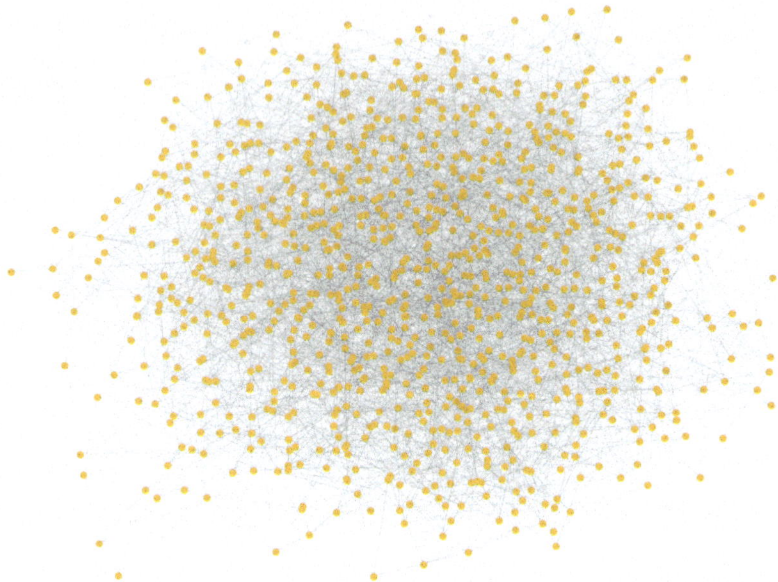

Fig. 3 Small-world network. NWS small-world network model with $v = 1000$ vertices, each of which is joined with its $k = 6$ nearest neighbors in the ring topology, and $p = 0.6$ probability of rewiring each edge

step 3 *Preferential attachment (PA)*. Each edge of u is attached to an existing vertex w with the following probability:

$$P_i = \frac{D(w)}{\sum_{u \in V} D(u)}$$

The defined probability is proportional to the degree of vertex u.

Holme and Kim [22] proposed a SF network model with two main characteristics: a perfect power-law degree distribution and a high clustering. To incorporate the second one, which is a peculiarity of the SW model, the authors modified the above BA algorithm by adding the following step:

step 4 *Triad formation (TF)*. If an edge (u, w) was added in the PA step, an edge from u to a neighbor of w (chosen randomly) is added. If all neighbors of w were already connected to u (i.e. there are no pair to connect), a PA step is done instead.

In this paper, we used the BA model with the fourth extra step to generate scale-free networks. In each experiment, we have chosen the number of vertices v between $1,000$ (see Fig. 4) and $15,000$, we add 5 random edges for each new vertex u, and we have chosen a probability $p = 0.3$ of adding a triangle after we have added each of these random edge.

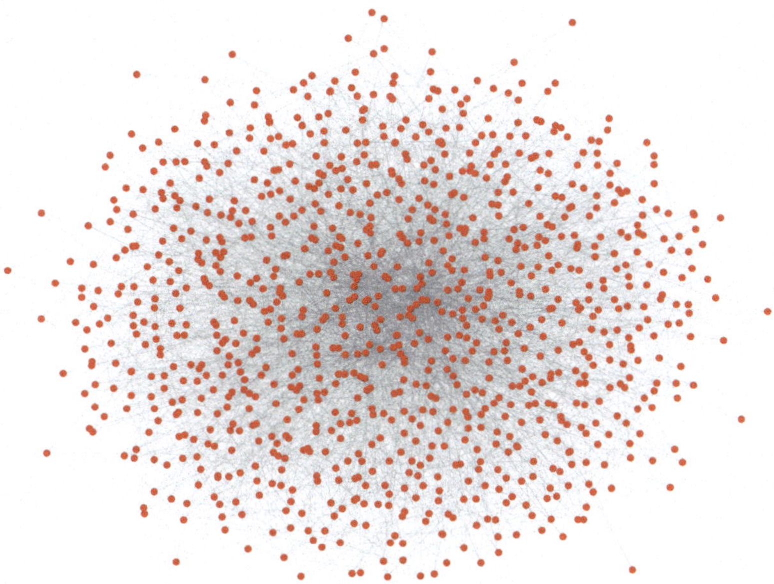

Fig. 4 **Scale-free network**. Extended BA model by Holme and Kim with $v = 1000$ vertices, $e = 5$ random edges to add for each new vertex and $p = 0.3$ probability of adding a triangle after adding a new edge

Real networks. In this paper we used three real networks: Dolphins, High Energy and Internet. The corresponding datasets have been downloaded from Mark Newman's website.

The *Dolphins* social network is an undirected and unweighted network of the relationships between the bottlenose dolphins (genus Tursiops) living in a community in New Zealand [25]. The dolphins have been observed between 1994 and 2001. This network is composed of 62 vertices which are the bottlenose dolphins and 159 edges which are the frequent associations (see Fig. 5a).

The *High Energy* theory collaborations is an undirected and weighted network of co-authorships between scientists who posted preprints on the High-Energy Theory E-Print Archive between January 1, 1995 and December 31, 1999 [28]. This network is composed of $8,361$ vertices which are scientists and $15,751$ edges which are connections existing if the scientists have authored a paper together (see Fig. 5b).

The *Internet* network was created by Mark Newman from data for July 22, 2006 and is not previously published. It was reconstructed from BGP tables posted by the University of Oregon Route Views Project. This network is a snapshot of the structure of the Internet at the level of autonomous systems (AS), i.e. collections of connected IP routing prefixes controlled by independent network operators. It is an undirected and unweighted network in which the vertices are $22,963$ AS and the edges are $48,436$ connections between AS (see Fig. 5c).

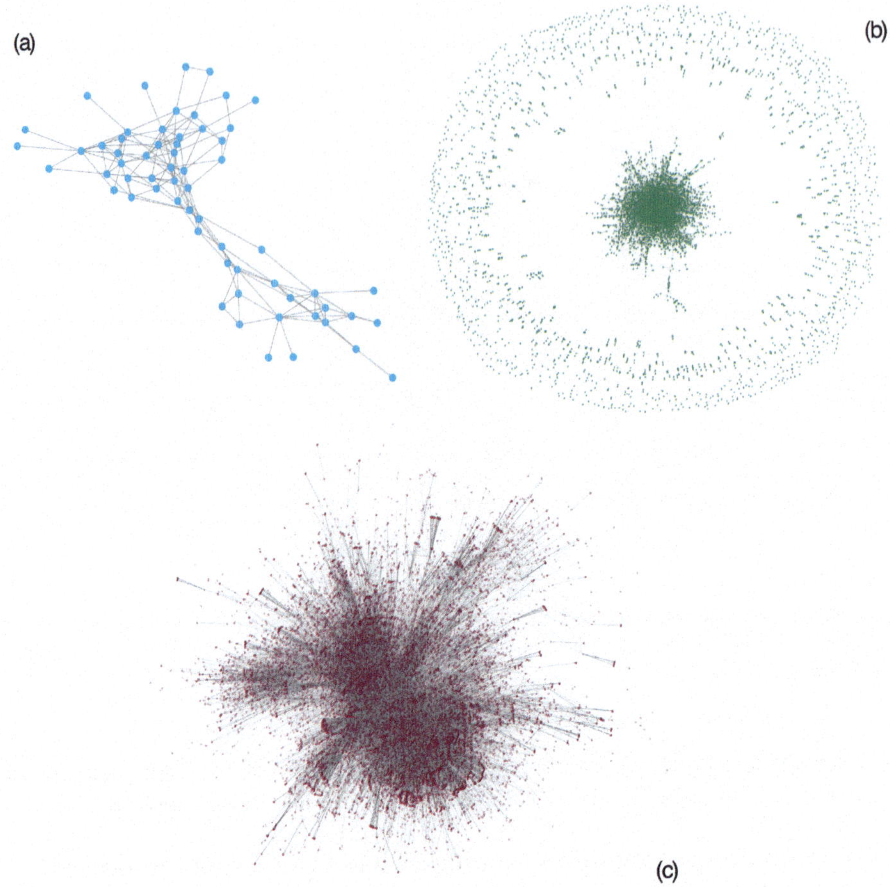

Fig. 5 Real networks. **a** *Dolphins* social network with 62 vertices (i.e. dolphins) and 159 edges (i.e. frequent associations). **b** *High Energy* theory collaborations network with 8, 361 vertices (i.e. scientists) and 15, 751 edges (i.e. connections existing if the scientists have authored a paper together). **c** *Internet* network with 22, 963 vertices (i.e. autonomous systems - AS) and 48, 436 edges (i.e. connections between AS)

4 Correlation Analysis

We investigated the correlations among the centrality measures introduced in Sect. 3.1, in both artificial and real-world networks described in Sect. 3.4. The network models include the SF networks, the SW networks and the ER random networks. For each class, we randomly generated five unweighted networks. Each network had between 1, 000 and 15, 000 vertices. Each SF network had between 4, 970 and 74, 959 edges. Each SW network had between 4, 810 and 71, 826 edges. Each ER network had between 5028 and 1, 125, 545 edges. The real-world networks

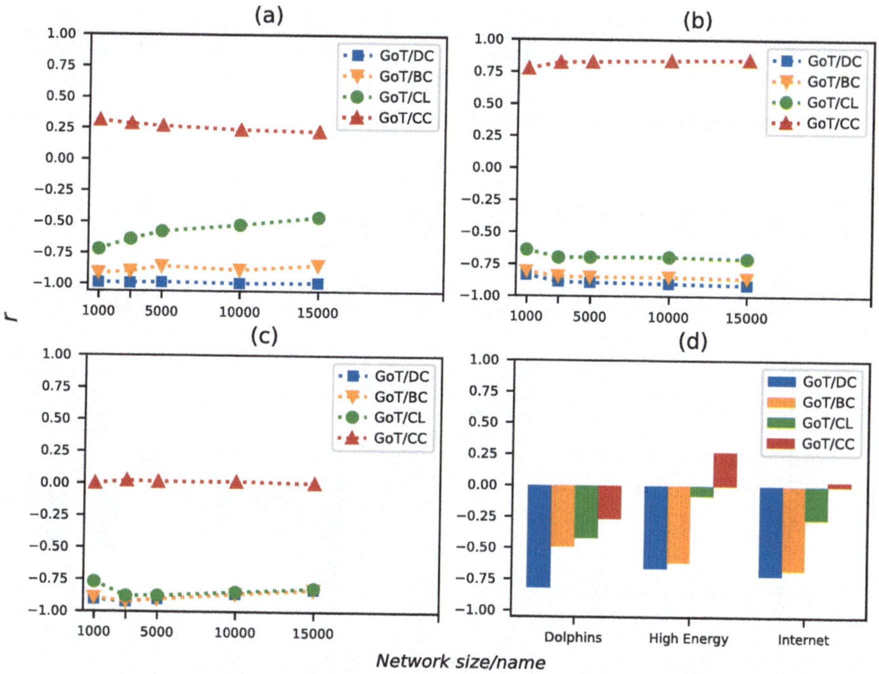

Fig. 6 Pearson correlation coefficient r between GoT and Degree (Blue), GoT and Betweenness (Orange), GoT and Closeness (Green), GoT and Clustering Coefficient (Red) as a function of the *Network size*, in SF networks (**a**), SW networks (**b**), ER random graph (**c**) and as a bar chart for real networks (**d**). In the artificial networks, the size is between 1000 and 15000 vertices

include three networks from different domains: the *Dolphins* social network, the *High Energy* theory collaborations and the *Internet* network.

For the implementation of the centrality measures such as degree, closeness, betweenness, and clustering coefficient, we used Python and NetworkX library [21]. For GoT we used the implementation by D.C. Mocanu [26] which is available on GitHub (https://github.com/dcmocanu/centrality-metrics-complex-networks), setting 1 thief and $\Phi_0^v = |V|$ vdiamonds per vertex. We let GoT to run for $T = \log^3|V|$ epochs. NetworkX was also used to generate the artificial networks and to perform our experiments with the real networks.

The results of the *Pearson correlation coefficient* r are presented in Fig. 6, the *Spearman rank correlation coefficient* ρ in Fig. 7 and the *Kendall Rank Correlation coefficient* τ in Fig. 8, with the growth of networks' sizes. Small deviations of rank correlation coefficients can be observed when the size of the networks is rather small. However, when networks grow big enough, the deviations are not visible anymore, especially for the rank correlation coefficients. Spearman correlation coefficient ρ was much higher than Pearson correlation coefficient r and so more capable of capturing the underlying ranking correlation between GoT and the other measures.

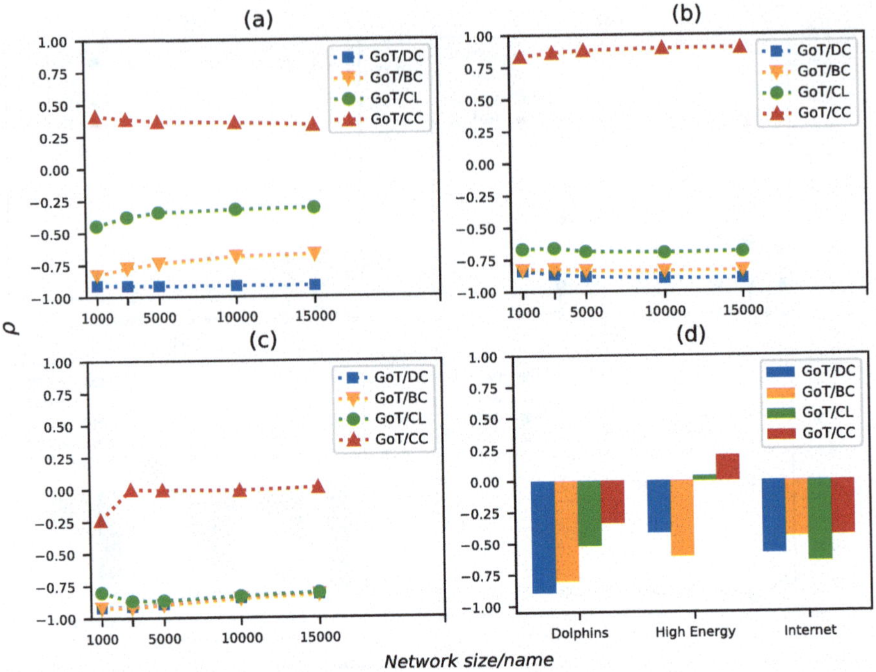

Fig. 7 **Spearman rank correlation coefficient** ρ between GoT and Degree (Blue), GoT and Betweenness (Orange), GoT and Closeness (Green), GoT and Clustering Coefficient (Red) as a function of the *Network size*, in SF networks (**a**), SW networks (**b**), ER random graph (**c**) and as a bar chart for real networks (**d**). In the artificial networks, the size is between 1000 and 15000 vertices

Moreover, we can observe that ρ is always larger than τ, but there is no distribution difference between these two coefficients.

GoT and degree centrality have the strongest negative correlation. GoT and betweenness centrality also exhibit a large negative correlation. GoT and closeness centrality are negative correlated, but this correlation is less than that between GoT and both degree and betweenness. GoT and clustering coefficient centrality have no correlation in most cases. In ER networks, we can observe the strongest and almost identical negative correlation among GoT and degree, betweenness and closeness. In SW networks, we can observe a very strong and unique positive correlation between GoT and the clustering coefficient. Real networks are more complex than the artificial ones, but also in this case the correlation among GoT and degree centrality is confirmed to be the strongest one.

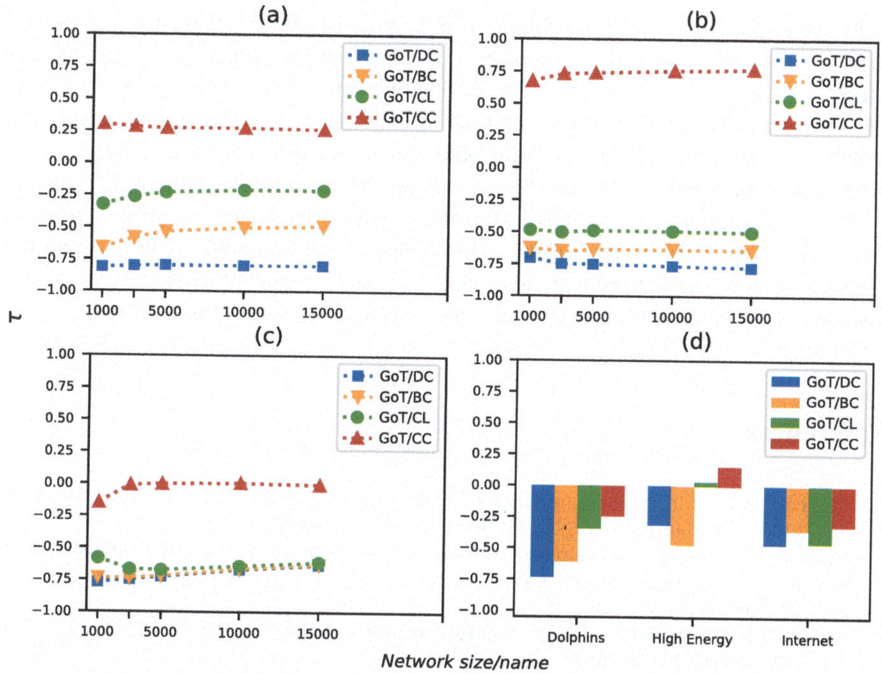

Fig. 8 **Kendall rank correlation coefficient** τ between GoT and Degree (Blue), GoT and Betweenness (Orange), GoT and Closeness (Green), GoT and Clustering Coefficient (Red) as a function of the *Network size*, in SF networks (**a**), SW networks (**b**), ER random graph (**c**) and as a bar chart for real networks (**d**). In the artificial networks, the size is between 1000 and 15000 vertices

5 Conclusions

In this work we examined the correlation between well known and recently proposed centrality measures in real and artificial networks, i.e. scale-free, small-world and Erdös-Rényi networks. If two centrality measures have a strong correlation, it means there is the possibility of approximating the metric with the highest computational complexity using the other. We used the Pearson correlation coefficient, the Spearman and Kendall rank correlation coefficients to study the correlations between the centrality metrics. An important finding is that the degree and the betweenness are strongly correlated with the new metric Game of Thieves. Also the closeness centrality is correlated with GoT but it's some sort of weaker correlation. The clustering coefficient and the Game of Thieves have a strong positive correlation only in SW networks.

We have done a correlation analysis observing the correlation coefficients when the number of vertices in both artificial and real networks increases. As future work, we want to make an analysis on artificial networks taking into account the increase of the number of edges when the number of vertices does not change.

Moreover, in this work, we focus on measures of vertex centrality and consequently on GoT's capability to compute the vertices centrality in a network. As future work, we want to apply the GoT algorithm to the case of edge centrality making a correlation analysis with the state-of-the-art measures of edge centrality. The centrality of an edge reflects its contribute spreading messages over a network, as short as possible, and we can use it as a tool for the community detection [7–9].

We can conclude that the GoT algorithm represents a step forward compared to the classical centrality algorithms which have at least a quadratic computational complexity an it can be used instead of degree, betweenness and closeness centrality when we want to compute the centrality of a vertex in a very large network.

References

1. Barabási, A.L., Albert, R.: Emergence of scaling in random networks. Science **286**(5439), 509–512 (1999). https://doi.org/10.1126/science.286.5439.509.
2. Bolland, J.M.: Sorting out centrality: an analysis of the performance of four centrality models in real and simulated networks. Soc. Netw. **10**(3), 233–253 (1988). https://doi.org/10.1016/0378-8733(88)90014-7.
3. Bonacich, P.: Power and centrality: a family of measures. Am. J. Soc. **92**(5), 1170–1182 (1987). https://doi.org/10.1086/228631
4. Brandes, U.: On variants of shortest-path betweenness centrality and their generic computation. Soc. Netw. **30**(2), 136–145 (2008). https://doi.org/10.1016/j.socnet.2007.11.001.
5. Calderoni, F., Catanese, S., De Meo, P., Ficara, A., Fiumara, G.: Robust link prediction in criminal networks: a case study of the Sicilian Mafia. Expert Syst. Appl. **161**, 113666 (2020). https://doi.org/10.1016/j.eswa.2020.113666.
6. Chen, P., Popovich, P.: Correlation: parametric and nonparametric measures. Sage university papers series. No. 07-139. Sage Publications (2002). https://books.google.it/books?id=UN4nAQAAIAAJ
7. De Meo, P., Ferrara, E., Fiumara, G., Provetti, A.: Enhancing community detection using a network weighting strategy. Inf. Sci. **222**, 648–668 (2013). https://doi.org/10.1016/j.ins.2012.08.001.
8. De Meo, P., Ferrara, E., Fiumara, G., Provetti, A.: Mixing local and global information for community detection in large networks. J. Comput. Syst. Sci. **80**(1), 72–87 (2014). https://doi.org/10.1016/j.jcss.2013.03.012.
9. De Meo, P., Ferrara, E., Fiumara, G., Ricciardello, A.: A novel measure of edge centrality in social networks. Knowl. Based Syst. **30**, 136–150 (2012). https://doi.org/10.1016/j.knosys.2012.01.007.
10. Duan, Y., Fu, X., Li, W., Zhang, Y., Fortino, G.: Evolution of scale-free wireless sensor networks with feature of small-world networks. Complexity **2017**, 1–15 (2017). https://doi.org/10.1155/2017/2516742
11. Erdös, P., Rényi, A.: On random graphs i. Publicationes Mathematicae Debrecen **6**, 290 (1959)
12. Faust, K.: Centrality in affiliation networks. Soc. Netw. **19**(2), 157–191 (1997). https://doi.org/10.1016/S0378-8733(96)00300-0.
13. Ficara, A., Cavallaro, L., De Meo, P., Fiumara, G., Catanese, S., Bagdasar, O., Liotta, A.: Social network analysis of Sicilian Mafia interconnections. In: Cherifi, H., Gaito, S., Mendes, J.F., Moro, E., Rocha, L.M. (eds.) Complex Networks and Their Applications VIII, pp. 440–450. Springer International Publishing, Cham (2020). https://doi.org/10.1007/978-3-030-36683-4_36

14. Fortino, G., Liotta, A., Messina, F., Rosaci, D., Sarne, G.M.L.: Evaluating group formation in virtual communities. IEEE/CAA J. Autom. Sin. **7**(4), 1003–1015 (2020). https://doi.org/10.1109/JAS.2020.1003237
15. Freeman, L.C.: Centrality in social networks conceptual clarification. Soc. Netw. **1**(3), 215–239 (1978). https://doi.org/10.1016/0378-8733(78)90021-7.
16. Friedkin, N.E.: Theoretical foundations for centrality measures. Am. J. Soc. **96**(6), 1478–1504 (1991). https://doi.org/10.1086/229694
17. Galaskiewicz, J.: Social Organization of an Urban Grants Economy: A Study of Business Philanthropy and Non-Profit Organizations. Academic Press (1985). https://books.google.it/books?id=Vd25AAAAIAAJ
18. Gilbert, E.N.: Random graphs. Ann. Math. Statist. **30**(4), 1141–1144 (1959). https://doi.org/10.1214/aoms/1177706098
19. Gkoulalas-Divanis, A., Labbi, A.: Large-Scale Data Analytics. SpringerLink : Bücher. Springer, New York (2014). https://books.google.it/books?id=1F24BAAAQBAJ
20. Grando, F., Noble, D., Lamb, L.C.: An analysis of centrality measures for complex and social networks. In: 2016 IEEE Global Communications Conference (GLOBECOM), pp. 1–6 (2016). https://doi.org/10.1109/GLOCOM.2016.7841580
21. Hagberg, A.A., Schult, D.A., Swart, P.J.: Exploring network structure, dynamics, and function using networkx. In: Varoquaux, G., Vaught, T., Millman J. (eds.) Proceedings of the 7th Python in Science Conference, pp. 11–15. Pasadena, CA USA (2008)
22. Holme, P., Kim, B.J.: Growing scale-free networks with tunable clustering. Phys. Rev. E **65**, 026107 (2002). https://doi.org/10.1103/PhysRevE.65.026107
23. Kendall, M., Gibbons, J.: Rank Correlation Methods. Charles Griffin Book. E. Arnold (1990). https://books.google.it/books?id=ly4nAQAAIAAJ
24. Li, C., Li, Q., Van Mieghem, P., Stanley, H.E., Wang, H.: Correlation between centrality metrics and their application to the opinion model. Eur. Phys. J. B **88**(3), 65 (2015). https://doi.org/10.1140/epjb/e2015-50671-y
25. Lusseau, D., Schneider, K., Boisseau, O.J., Haase, P., Slooten, E., Dawson, S.M.: The bottlenose dolphin community of Doubtful Sound features a large proportion of long-lasting associations. Behav. Ecol. Sociobiol. **54**(4), 396–405 (2003). https://doi.org/10.1007/s00265-003-0651-y
26. Mocanu, D.C., Exarchakos, G., Liotta, A.: Decentralized dynamic understanding of hidden relations in complex networks. Sci. Rep. **8**(1), 1571 (2018). https://doi.org/10.1038/s41598-018-19356-4
27. Newman, M., Watts, D.: Renormalization group analysis of the small-world network model. Phys. Lett. A **263**(4), 341–346 (1999). https://doi.org/10.1016/S0375-9601(99)00757-4.
28. Newman, M.E.J.: The structure of scientific collaboration networks. Proc. Natl. Acad. Sci. **98**(2), 404–409 (2001). https://doi.org/10.1073/pnas.98.2.404.
29. Oldham, S., Fulcher, B., Parkes, L., Arnatkeviciute, A., Suo, C., Fornito, A.: Consistency and differences between centrality measures across distinct classes of networks. PLOS ONE **14**(7), 1–23 (2019). https://doi.org/10.1371/journal.pone.0220061
30. Pace, P., Fortino, G., Zhang, Y., Liotta, A.: Intelligence at the edge of complex networks: the case of cognitive transmission power control. IEEE Wirel. Commun. **26**(3), 97–103 (2019). https://doi.org/10.1109/MWC.2019.1800354
31. Ronqui, J.R.F., Travieso, G.: Analyzing complex networks through correlations in centrality measurements. J. Stat. Mech. Theory Exp. **2015**(5), P05030 (2015). https://doi.org/10.1088/1742-5468/2015/05/p05030
32. Rothenberg, R.B., Potterat, J.J., Woodhouse, D.E., Darrow, W.W., Muth, S.Q., Klovdahl, A.S.: Choosing a centrality measure: Epidemiologic correlates in the Colorado Springs study of social networks. Soc. Netw. **17**(3), 273–297 (1995). https://doi.org/10.1016/0378-8733(95)00267-R.
33. Scott, J.: Social Network Analysis: A Handbook. SAGE Publications (2000). https://books.google.it/books?id=Ww3_bKcz6kgC
34. Shao, C., Cui, P., Xun, P., Peng, Y., Jiang, X.: Rank correlation between centrality metrics in complex networks: an empirical study. Open Phys. **16**(1), 1009–1023 (2018). https://doi.org/10.1515/phys-2018-0122

35. Spearman, C.: General intelligence, objectively determined and measured. Am. J. Psychol. **15**(2), 201–292 (1904). https://doi.org/10.2307/1412107.
36. Stephenson, K., Zelen, M.: Rethinking centrality: methods and examples. Soc. Netw. **11**(1), 1–37 (1989). https://doi.org/10.1016/0378-8733(89)90016-6.
37. Valente, T.W., Coronges, K., Lakon, C., Costenbader, E.: How correlated are network centrality measures? Connections (Toronto, Ont.) **28**(1), 16–26 (2008). https://pubmed.ncbi.nlm.nih.gov/20505784
38. Valente, T.W., Foreman, R.K.: Integration and radiality: measuring the extent of an individual's connectedness and reachability in a network. Soc. Netw. **20**(1), 89–105 (1998). https://doi.org/10.1016/S0378-8733(97)00007-5.
39. Wasserman, S., Faust, K., Granovetter, M., Iacobucci, D.: Social Network Analysis: Methods and Applications. Structural Analysis in the Social Sciences. Cambridge University Press, Cambridge (1994). https://books.google.it/books?id=CAm2DpIqRUIC
40. Watts, D.J., Strogatz, S.H.: Collective dynamics of 'small-world' networks. Nature **393**(6684), 440–442 (1998). https://doi.org/10.1038/30918

A LPWAN Case Study for Asset Tracking

Fabrizio Formosa, Michele Malgeri, and Marco Vigo

Abstract The industrial application of Internet Of Thing is rapidly growing leading to, so called, Industrial IoT. Moreover, the future spread of 5G make possible the application to many new fields. However, waiting for 5G, Low-Power Wide-Area Network (LPWAN) technologies permits company to already develop applications energy-efficient using current communication technologies. This paper presents the development of a project mainly aiming at providing tracking service starting from LPWAN protocol selection till the analysis of some implementation prototypes. The work focuses on energy saving when using GPS and cryptography routines.

1 Introduction

Internet Of Things is one of the most interesting industrial sector, it can be considered a mature technology with a growing number of cases already in use covering a multitude of options and facing problems dealing, for instance, with connectivity, hardware/software and energy saving.

Today, more than one-third of companies use IoT solutions so that we use the acronym IIoT (Industrial IoT) to refer to the use of IoT principles in industry. With the expansion of IoT, Low-Power Wide-Area Network (LPWAN) technologies are increasingly used in industrial and research communities for their low power, long range and low-cost requirements. Sigfox, LoRaWAN and Narrowband-IoT (NB-IoT) are the leading LPWAN technologies that differ in different technical aspects [13]. Low Area Wide Network (LPWAN) is a category of wireless communication technologies. These technologies are used in the IoT field to connect a large number of

F. Formosa · M. Malgeri (✉) · M. Vigo
Dip. Ingegneria Elettrica Elettronica Informatica, Universitá degli Studi di Catania, Viale A. Doria 6, Catania, Italy
e-mail: michele.malgeri@dieei.unict.it

F. Formosa
e-mail: fabriformo@gmail.com

M. Vigo
e-mail: vigomarco@live.it

G. Fortino et al. (eds.), *Data Science and Internet of Things*, Internet of Things,
https://doi.org/10.1007/978-3-030-67197-6_4

"things" such as smart water meters, parking sensors, GPS locators and countless other objects [19]. All LPWAN technologies have three fundamental characteristics in common: low power, long range and low-cost requirements.

The LPWAN, and in particular NB-IoT, in a way, aims at anticipating 5G. Mobile cellular networks such as 3G, 4G and 4.5G ensure high bandwidth and extended coverage for several kilometers in the face of high connection costs and high energy consumption. To partially overcome these limitations while waiting for 5G, many companies, in recent years, opted for NB-IoT as IoT and M2M connectivity technology. NB-IoT is a versatile and very economic technology that allows companies to create IoT applications in the most varied areas: Smart Agriculture, Smart Metering, integrated logistics, digital healthcare.

The paper presents a case study dealing with tracking where resources are low, mainly in energy.

It consists of a Smart application aiming at tracking livestocks using an LPWAN network, and in particular Narrowband IoT [6] that is a communication protocol based on LTE protocol [17]. The work presents the advantages in choosing NB-IoT among the other technologies already available. The case study aims at highlighting some specific implementation problems regarding the impact of enabling security to energy management in contexts where the use of GPS is critical for consuming reasoning.

The system is based on EVB-M1 development board [9] supplied by Huawei with the expansion board containing a GPS module and a SIM card that allowed us to connect to a working NB-IoT network.

Several tests were carried to study the battery evolution mainly during the use of NB-IoT network, and GPS. The tests were carried out, also, to study the impact encryption.

Next section presents a brief presentation of some IIoT platforms useful for tracking application, than Sect. 3 discusses the problem that we aim at solving, Sect. 4 presents the actual implementation of the platform in deepening problems, mainly in energy saving, and solutions and in Sect. 5 some evaluation of the platform behaviour in terms of functionality and perform are discussed.

2 State of the Art of Platform for IIOT

The main LPWAN technologies developed in recent times are Sigfox, NB-IoT and LoRa® and they are likewise quite different.

Sigfox technology [18] was founded in 2010 by Ludovic Le Moan and Christophe Fourtet with the aim of connecting objects and is present in over 70 countries. Sigfox uses a free frequency spectrum (without a license) and the number of messages over the uplink is limited. LoRa® is a network technology developed by Cycleo, founder of LoRa Alliance®.

LoRaWAN® is the open standard to connect wireless "things" and is designed to scale from a single gateway installation up to large global networks with billions of devices [12]. LoRaWAN® uses an unlicensed frequency spectrum, also, and requires a gateway that receives the messages and forwards them to the router. NB-IoT is a cellular radio technology specified by 3GPP in Release 13 for wide area and low power connectivity. In September 2019, Global mobile Suppliers Association announced that over 142 operators have deployed/launched either NB-IoT or LTE-M networks [5].

NB-IoT, unlike Sigfox and LoRa®, uses a licensed frequency spectrum thanks to which the interference is reduced. Now let's describe the main features of Sigfox, LoRaWAN and NB-IoT.

2.1 Sigfox

Sigfox™ uses the unlicensed sub-GHz ISM bands (868 MHz in Europe, 915 MHz in North America, and 433 MHz in Asia) and an Ultra-Narrow Band (UNB) modulation (100 Hz). It requires the use of proprietary base stations. The end-devices use BPSK modulation in the already cited UNB with a maximum data rate of 100 bps. Thanks to UNB in sub-GHz spectrum Sigfox™ has very low noise levels and makes efficient use of the frequency band. All this feature lead to very low power consumption, high receiver sensitivity, and low-cost antenna design. The maximum payload length for each uplink message is 12 bytes and the number of these messages is capped to 140 messages per day. Regarding the downlink messages, the maximum payload length for each one is 8 bytes and their number is limited to 4 messages per day, thus all uplink message could not be acknowledged.

Given the impossibility to use acknowledgments, the reliability of the uplink transmission is guaranteed by time/frequency diversity and transmission duplication. A generic end-device sends message three times through different frequency channels, all base stations can receive messages simultaneously over all channels. So, end-devices can transmit the message a channel randomly chosen. This mechanism reduces the end-device complexity and its cost.

2.2 LoRaWAN

LoRa® is a spread spectrum modulation technology that use the unlicensed sub-GHz band, it's derived from the chirp spread spectrum (CSS) modulations that guarantees a full bidirectional communication. The signal has a low level of noise, high interference resilience, and is difficult to detect or jam. LoRaWAN is a protocol that define the upper layers of the LoRa network, provides six spreading factors (SF7 to SF12) to adapt the data rate and the range of the transmission. A higher spreading factor means a longer transmission range and lowest data rate. LoRaWAN provides a data rate between 300 bps and 50 kbps and a maximum payload length for each message of 243 bytes.

A message transmitted by a LoRaWAN end-device is received by all base stations in the range. This multiple reception improves the communication reliability ratio; however, it is necessary to deploy many base stations in each area, increasing thus the network infrastructure cost.

LoRaWAN provides different communication classes to meet the different latency requirements of IoT applications:

- Class A (bidirectional end-devices): each uplink transmission is followed by two time-windows for receiving downlink messages. The uplink transmission time is chosen by the end-device according to its communication requirements. In this class belong the lowest power end-device systems for IoT applications, these systems require short downlink communication after the transmission of an uplink message. Downlink transmission will always have to wait until the next uplink message;
- Class B (bidirectional end-devices with scheduled receive slots): these devices support the receive windows of class A and in addition can open extra receive windows at scheduled times for receive downlink message in other instants of time. To open this extra window, end-devices receive a time synchronized Beacon from the base station, so the server can know when the device is listening.
- Class C (bidirectional end-devices with maximal receive slots): these end-devices have a *receive windows* constantly open. This class have the maximum energy consumption and is defined for IoT applications with continuous power supply.

2.3 NB-IoT

Narrowband IoT (NB-IoT) is an LPWAN technology with licensed bandwidth based on the LTE protocol. NB-IoT, in fact, uses a part of the LTE standard, but limits the bandwidth to a single band of 200 kHz. NB-IoT has been optimized for the transmission of small and infrequent data and significantly improves the power consumption of user devices, system capacity and spectrum efficiency, especially in deep coverage. New physical layer signals and channels allow rural and deep indoors coverage. The basic technology of NB-IoT is much simpler than that of GSM/GPRS and its cost is expected to decrease rapidly as demand increases [6].

As in LTE, NB-IoT uses OFDM modulation for downlink communication and SC-FDMA for uplink communication [17]. NB-IoT can be deployed in 3 different modes: In-band, Guard-band and Stand-Alone (Fig. 1).

- In-Band deployment, through the use of one or more portions of the 180 kHz spectrum allocated in the useful LTE band with Physical Resource Blocks (PRB) to guarantee the use of the band only to some terminals;
- Guard-Band deployment, through the use of one or more 180 kHz PRB allocated in the LTE guard band;
- Stand-alone deployment, through the use of one or more channels of 200 kHz nominal, 180 kHz effective [20].

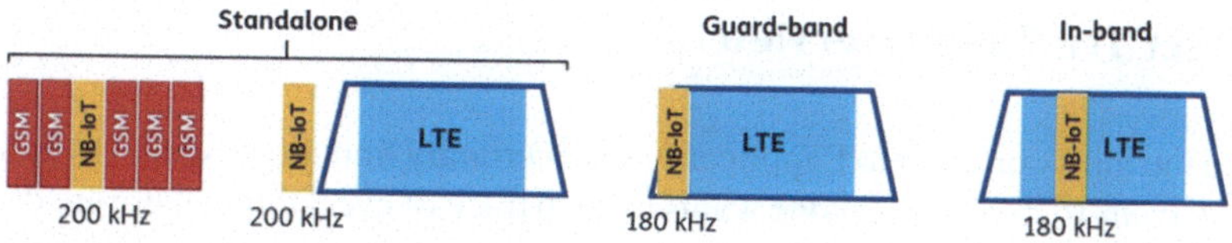

Fig. 1 Operation modes for NB-IoT [20]

NB-IoT allows to connect more than 100 K devices per cell and it could be increased by exploiting multiple NB-IoT carriers. The maximum data rate used by NB-IoT is 200 kbps in downlink and 20 kbps in uplink, with 1600 bytes of payload size in each message. The battery life of a device that uses NB-IoT technology can reach 10 years by transmitting an average of 200 bytes per day [13].

2.3.1 LoRaWAN Versus NB-IoT

The benefits of NB-IoT standard include: the ability to deploy network based on existing LTE networks, a large coverage area, and high quality of customer service. Advantages of the LoRaWAN standard are low cost of network deployment and user terminals, large coverage area, long service life of user terminals, high penetrating power, but relatively low achievable data rates.

The cost of deploying the networks is one of the most important and critical factors to the success. The most expansive solution is NB-IoT, because this standard requires the modernization of existing base stations, the cost of which starts from 15000 €. The cost of base stations for LoRaWAN is much smaller since it varies from 100 € to 1000 €.

According to [10], the implementation of LPWAN standards operating in non-licensed bands (LoRa) faces great difficulties due to the limitations on power and dedicated bands. The maximum range for LoRa is indicated for the radiated power of 500 mW, which is the maximum allowed in Europe. Technology NB-IoT works at frequencies assigned to Telecom operators, so these limitations do not affect it.

Table 1 Overview of LoRaWAN and NB-IoT

Feature	LoRaWAN	NB-IoT
Modulation	CSS	QPSK
Frequency	Unlicensed ISM bands	Licensed LTE bands
Bandwidth	250 and 125 kHz	200 kHz
Maximum data rate 650 kbps	200 kbps	
Maximum messages/day	Unlimited	Unlimited
Maximum payload length	243 bytes	1600 bytes
Range	5 km (urban), 20 km (rural)	1 km (urban), 10 km (rural)
Authentication and encryption	Yes (AES 128b)	Yes (LTE encryption)

3 Asset Tracking Overview

Among the most common IoT applications we certainly find those related to the *asset tracking context*. Today, everyone wants to keep track of everything, there are devices to track packages, to track cars as well as devices tracking animals, up to devices that track people themselves, and in a scenario such like this, IoT plays a significant role. Usually, all these problems belongs to *Asset Tracking*. Asset tracking means all related to the method of tracking the physical assets, for instance by the scan of barcode labels bound to the assets or through tags with GPS, BLE or RFID system that transmit their location. These technologies are used also for indoor tracking of persons, often wearing a specific tag.

There are many use cases that belong to the asset tracking, and almost all of them concern the IoT scenario. We have classified them based on some key features, that highlight the economic value of the tracked entity and the various aspects that can affect energy consumption.

- Entity tracked (item, vehicle, animal, person)
- Route (unknown, known, inside a specific area, from point A to point B)
- Position's sampling rate (high, low)
- Device power supply (constrained, unlimited)

Analysing the use cases following these characteristics, it emerges that in many of them the use of LPWAN plays a key role in their implementation. For example, in the use case related to the GPS track of livestock, a multi-year battery is required, also there are high cost constraints, so LPWAN technologies with their low power consumption represent a suitable choice. In the use case of the GPS track of objects (e.g.: shipments), it may be more convenient to track the trailers instead of the trucks: while trucks were tracked by powered cellular trackers, trailers do not have power supply and require battery powered trackers and therefore LPWAN is again the right technology [16].

In the following subsection we present some of the most common Use Cases organized in base of the Entity Tracked.

3.1 Vehicles

Use cases related to Vehicles include several applications. An interesting example is bus tracking as anti-theft system. Bus tracking systems often works using a smartphone application and a panel displaying the location after a specific time interval [4]. In the most recent implementations, the use of the LoRaWaN network is proposed to reduce maintenance costs [7], since LPWANs, as already mentioned, have significantly reduced costs compared to the use of other technologies.

The use case related to theft detection is perhaps one of the most developed. In general, systems of this type are not affected by the problem of energy consumption, as they exploit the power supplied by the vehicle's battery. In some implementations the theft detection is simply signalled by external apps, in others driver authentication can be requested and countermeasures can be applied directly on the vehicle in question [21].

Other common applications based on a GPS tracking are the detection of vehicles that exceed a certain speed limit and assurance motoring of car accident.

3.2 *Objects*

Speaking of objects in general, we can divide them into high value objects and low value objects. As use cases related to low value objects, we present rental bicycle tracking, pallet tracking, trailer tracking and laptop tracking.

In the first one the GPS position is transmitted only in the event of loss, using the search function via an external device or in the event of violation of a specific perimeter. Some application does not use GPS when possible in order to last more and increase coverage [1, 2].

In the use case related to pallet, GPS tracking can monitor their route, sending notification when the pallets are at the retail store. It can also monitor the conditions of transportation of the goods on top of the pallet, and finally using LPWAN technologies it's possible to connect them to reduce their loss [16].

When you want to track objects carried by trucks, could be more convenient to trace the trailer rather than the objects in question or the entire vehicle. Not tracking the vehicle resolves several problems related to privacy. Tracking the trailer does not require any other operations when the truck transporting it is replaced or when it is left at the depot.

Finally, we can use theft detection system to track a stolen laptop through the implementation of GPS, GSM, Motion Sensor, and Cloud Services. This solution uses an external system that doesn't require the laptop to be switched on [3].

Regarding high value objects (for example: dangerous objects, medicines, precious stones like diamonds, currency, etc.) In this case we need a very reliable system, and given the nature of the traced entity, we must not have power supply problems, and continuous monitoring is required. In this context, the cost of the tracking device is not relevant.

3.3 *People*

One of the application scenarios in this area is to monitor the position of the elderly, for example by providing them with a device that also allows them to ask for help if they get hurt. Another scenario is to monitor children, for example when they go to school. Both mentioned solutions are available on the market. Privacy is a big concerns when tracking peoples, a lot of questions arise, for instance, *where data will be stored? What laws am I supposed to follow? Is it possible to use metadata?*

Another application scenario is related to tracking and monitoring patients with mental disorder. Patients with mental disorder can lose situational awareness when they are in public places, and this can involve very high risks. The idea is to develop a device that allows the psychiatrist to monitor the position and status of the patient using a GPS system and a LoRaWAN communication system [15]. In this case the use of LoRaWAN technology offers a longer battery life and a more efficient implementation thanks to the capability to operate at low power that means more reliability and scalability.

3.4 Animals

In the case of animal tracking it is possible to distinguish two large families of applications. The first, much simpler, concerns the GPS tracking of pets, with small devices that can be installed in their collar. These types of devices are already on the market and can be purchased.

The other type concerns the monitoring of livestock. The applications of this use case have already been designed and tested on a 3G transmission technology [11], however a technology like LPWAN, and in particular NB-IoT, is more adequate solution to satisfy the requirements because it is the only LPWANs technology that uses a licensed band. In fact, despite involving a higher cost, due to more expensive modem and to the SIM used to transmit, NB-IoT the cost is justified in the case of entities of such high value (like livestock). Furthermore, the signal based on NB-IoT network has strong penetrability and great advantage in coverage with respect to the other similar technologies, which enhance reliability and security of data transmission.

4 Platform Description

The proposed application is based on an NB-IoT development kit supplied by IoT Club [9]. This kit includes a Huawei evaluation board and recommends the use of the LiteOS operating system as it is more optimized for creating applications on constrained resources devices. The kit also provides a step-by-step guide for the development of applications on LiteOS completed by an example source code that can be imported from GitHub [8].

The platform architecture is pictured in Fig. 2 that shows the main components:

- Evaluation board Huawei EVB-M1: it controls the operations of data acquisition, data transformation and data transmission
- GPS expansion board: it allows the acquisition of GPS coordinates;

Fig. 2 System architecture

- Modem Quectel BC35-G: it allows the transmission of data and diagnostic information on NB-IoT network.

In the following, it is given a description of the developed system, beginning from the Hardware components up to the Software programming part.

To enable the communication between the board and the modem, we used AT commands and as first step we set the parameters needed to enable the device transmission and to connect it to the network.

In order to reduce energy consumption a functionality is provided, based on AT commands. In-fact, it is possible to put the modem in Power Safe Mode by turning off the radio for a certain amount of time. When the data is transmitted, the radio is automatically switched on. Is important to emphasize that through some AT commands it is possible to activate automatic diagnostic information by Universal Remote Console framework (ISO/IEC 24752-1:2014).

For implementing the power safe mode of modem, we need to define two different timers, that will be used for switching between the power safe states.

- Active time: the time after that the modem switch in the state with the lowest energy consumption
- Periodic Tau: the time after that the modem "awakes" from the lowest energy consumption state

A switched-on NB-IoT communication device can take three different states: Connected, Active, Deep-sleep (Fig. 3).

- Connected: the modem transmits or receives data on the network. In this mode the energy consumption is maximum and if there are no transmission or reception messages, the modem automatically switches to Active state.
- Active: the modem is turned on and ready to communicate with the network but is not actually communicating anything. At regular intervals it does the "paging" to see if there are any messages received. At the end of a timer set by the user (Active Time), the modem automatically switches to Deep-Sleep state. In this state, energy consumption is average.

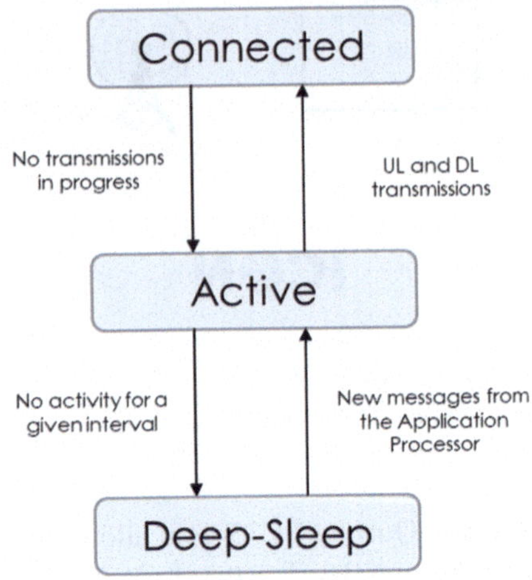

Fig. 3 FSM describing the states that the modem can assume

Table 2 Commands enabling Powersafe mode

AT+NPSMR	Readout of the modem power mode. If it is equal to 1, it enables the URC relating to power mode
AT+CPSMS	Enable power safe mode and set timers

- Deep-sleep: the modem is turned off and cannot be received data from the network. Only the UART serial interface is enabled to receive AT commands and energy consumption is minimal. At the end of another timer set by the user (Periodic TAU), the modem automatically switches to Active state without having to redo the connection phase.

Based on what we called *Powersafe Mode* must last at least

$$t_{powerSafe} = \tau - t_{active}$$

Table 2 lists the AT commands needed to enable the Powersafe mode and to set the timer to correct value.

4.1 GPS Data Acquisition

Module GPS L80-R is used to obtain GPS coordinates. The communication between this module and the board occurs through UART, and an initialization function is required for using it (Table 3).

Table 3 Commands 2

AT+NCDP	Sets IP address and port number related to the server where the data will be transmitted
AT+NBAND	Sets the modem operating band. Each band is identified by a specific number, based on the modem you are using. It is possible to set different bands
AT+COPS	Sets the PLMN (Public Land Mobile Network) of the telephone operator
AT+CDGCONT	Allows the definition of the connection parameters relating to the various network elements, including the APN (Access Point Name)
AT+CFUN=1	Activates the radio part of the modem. N.B. Before coming some parameters such as CDP and APN, it is necessary to set CFUN to 0 (by turning off the radio part) and then reset it to 1 once the configuration is complete
AT+CGATT=1	Registers the modem to the GPRS service
AT+CGPADDR	Checks the assignment of an IP address
AT+NCONFIG	Sets some modem parameters (es: AUTOCONNECT,TRUE)
AT+NRB	Restarts the modem. The parameters defined with “AT+NCONFIG” become effective after restart
AT+NSOCR=DGRAM,17,5682,1	Creates a socket
AT+NSOST	Used to transmit the data through a socket. Requires the data in hexadecimal format, IP address, port number, socket number and message length in bytes. If the transmission was successful, TRUE will be returned, otherwise FALSE

Data acquisition occurs by transmitting a specific command to the GPS module that look for current position; as already mentioned the module used supports the NMEA 0183 standard [14].

Each NMEA message (see Fig. 4 for an example) starts with ‘$’ and ends with ‘< CR > LF >’. ‘PMTK314’ indicates packet type and message type; ‘*’ indicates the end of the data field and the following number is the checksum. This message is used to set NMEA sentence output frequencies. There are totally 19 data fields that present output frequencies for the 19 supported NMEA sentences individually. In the case of figure the value ‘1’ in the second data field means that the module only output Recommended Minimum Position Data (RMC) once everyone position fix.

```
$PMTK314,0,1,0,0,0,0,0,0,0,0,0,0,0,0,0,0,0,0,0*29<CR><LF>
```

Fig. 4 NMEA message

```
$GPRMC,060946.000,A,3150.7815,N,11711.9239,E,2.87,314.13,050314,,,D*69<CR><LF>
```

Fig. 5 NMEA response

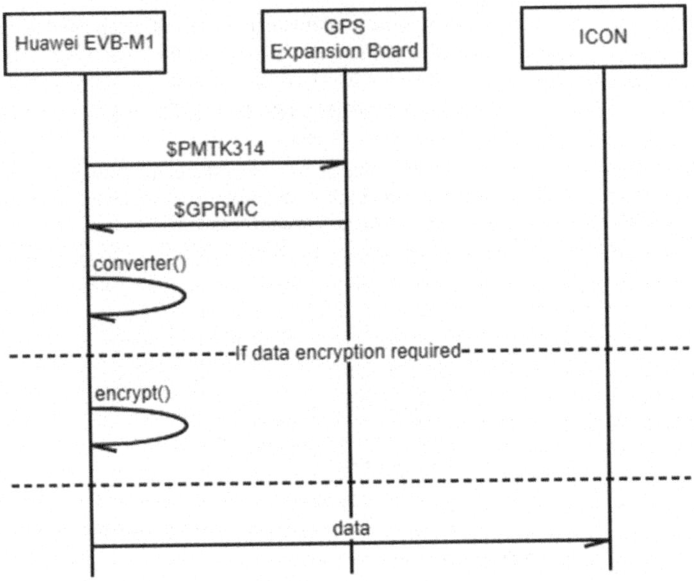

Fig. 6 Exchange of messages

The response has the format response (see Fig. 5 contain information about position, velocity, time.

The information about Latitude and Longitude, relevant for our application, is as follows:

- Latitude: ddmm.mmmm
- Longitude: dddmm.mmmm

Where "d" is degree and "m" is minutes. This information will be converted in the following decimal format through the function *converter()*:

- Latitude: dd ± (mm.mmmm/60)
- Longitude: ddd ± (mm.mmmm/60)

Where '±' depends on whether we are considering North or South Latitude, and East or West Longitude (Fig. 6).

The data obtained (e.g.: Lat: 37.527731; Lon: 15.075191) will finally be transmitted over the NB-IoT network using AT-commands and the CoAP protocol to the ICON Platform (Table 1).

4.2 AppBattery

Since one of the main objectives of this project is to monitor energy consumption in asset tracking GPS based IoT applications on constrained resources devices, it is essential to regularly notify the user with the charge level of the battery.

Unlikely, Huawei LiteOS does not provide a function to read the battery level since this function depends on the Board Support Package (BSP) platform. In embedded systems, a BSP is the layer of software containing hardware-specific drivers and other routines that allow a particular operating system (traditionally a real-time operating system, or RTOS) to work in a particular hardware environment (a computer or CPU card), integrated with the RTOS itself using. Usually, third-party hardware developers who wish to support a particular RTOS provide the layer together with hardware. In summary, BSP can be defined as the software layer that contains specific hardware drivers allowing the OS to operate in a particular hardware.

To be able to read the battery value, we connected the positive pole of the battery to a pin (PA7), using an external hardware. It is necessary to use the ADC to convert the analog value of the voltage read from pin PA7, to a digital number. Then, the value obtained must be re-scaled to a range coherent with the percentage of the battery. The function reads several times the value on the pin and then make the average of all these values read (this avoid possible error values). The obtained value is then reported in a range of values related to the voltage required to power the device. The data so obtained represent the app report which will be transmitted.

4.3 Encryption

The project include a 128-bit AES encryption message facility that allows to exchange over the network encrypted messages. Although the NB-IoT technology takes advantage of LTE network and its security against attacks it is better to provide encryption of messages in order to defend, for instance, against impersonation attacks. We used symmetric cipher using a pre-shared key. Despite ST provides specific libraries for the encryption algorithms, due to the evaluation board and the SDK of the operating system (both still in development), it was not possible to use these libraries within the project. Therefore, we used a generic C implementation of the AES algorithm with ECB encryption mode not optimized for MCU. The encryption process takes place directly within the transmission function (bc95_send), after the conversion to hexadecimal format, in order to provide the encryption of all report messages for each App.

5 Platform Evaluation

In this section we provide the reader with evaluation of cost and performance evaluation presenting some test results, also.

5.1 Cost and Performance Evaluation

To evaluate the efficiency of the system, we carried out an energy-consumption analysis taking as a reference the battery discharge process in three different tests. At this stage of developing process of the prototype it is not possible to activate the power saving mode of the MCU, therefore, in the various tests carried out, was implemented only the Power Safe Mode of the modem. By observing the results obtained and comparing the curves of the battery discharge process with or without encryption, it is possible to evaluate how these encryption operations impacts on energy consumption. The system under test is the complete system running two apps: AppGPS for tracking and AppBattery for monitoring power consumption. The system was installed in a car moving around to capture real positioning data. Each test differs in the following parameters:

- App Cycle: cycle values in the 2 running apps (how many seconds the app function is repeated);
- Power Safe Mode parameters of the modem: the values relating to the Power Safe Mode of the modem (Active Time and Periodic Tau);
- Using of encryption algorithm.

5.2 Experiments

This section describes the setup of tests done and the results. In the graphs the x-axis represents the voltage while y-axis is the time. In the graphs the green line indicates the minimum voltage required by the system for correct operation, it is important to emphasize that the board can continue to operate up to a minimum voltage of 2.8 V, but below 3.2 V it will no longer be possible to transmit the data. The red lines highlight 90% and 50% of the discharge process.

All the experiments were repeated several times and results represents the average behavior of the setups.

Table 4 Setup #1

Parameter	Value	Unit
AppGPS cycle	5	Minutes
AppBattery cycle	10	Minutes
Active time	10	Seconds
Periodic Tau	5	Minutes
Encryption	Yes	

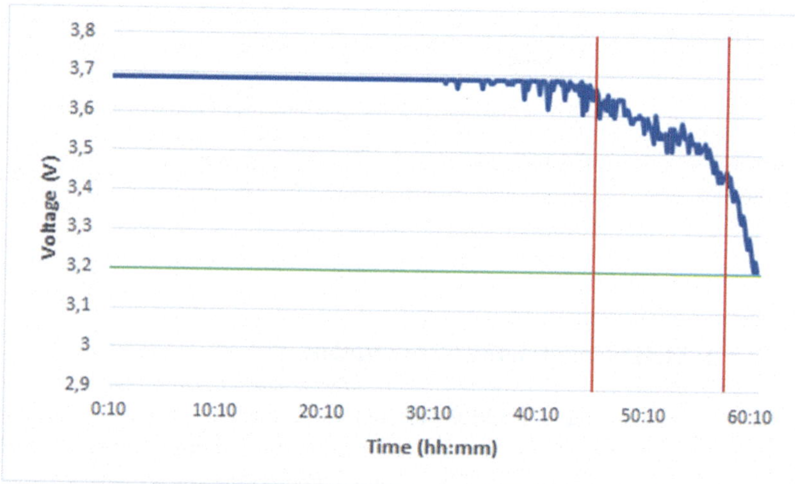

Fig. 7 Result of Test #1

5.2.1 Test 1: Stress Test

The Table 4 shows the meaningfully parameters of the first setup. These parameters represent a typical heavy condition, since the transmission of data is quite continuous. This test aims at evaluating the performance of the battery in *stress condition.*

In the average the battery is able to provide enough energy to support system up to almost 43 h. Figure 7 shows the discharge rating of the battery.

It is possible to note that there is a change in the trend of the curve once is passed the 90% of the charge, that signs the beginning of the discharge process. This process starts to accelerate reaching the 50% of the charge. It is possible to note that battery level remains at maximum level for 24 h.

Table 5 Setup #2

Parameter	Value	Unit
AppGPS cycle	12	Hours
AppBattery cycle	12.5	Hours
Active time	2	Seconds
Periodic Tau	310	Hours
Encryption	Yes	

Table 6 Setup #3 (without encryption)

Parameter	Value	Unit
AppGPS cycle	5	Minutes
AppBattery cycle	10	Seconds
Active time	10	Seconds
Periodic Tau	5	Minutes
Encryption	No	

5.2.2 Test 2: Minimizing Energy Consumption

The Table 5 shows the meaningfully parameters of the second setup. We choose these parameters for monitoring the system in a condition that minimize the energy consumption limiting the transmissions as much as possible.

Unlike the first test, it is possible to note that modifying the execution time of the various apps and above all the sleep times of the modem (which are maximized here), the battery had a much longer life, in-fact in this test the battery lasted for almost four days. The discharge process is very close to the previous case but the duration.

5.2.3 Test 3: No Encryption

The Table 6 shows the parameters of the third setup. These parameters are the same of the first setup, with encryption disabled. This test aims at comparing the impact of encryption on energy consumption.

The Fig. 8 shows the result of this test where the battery lasted for almost 61 h, i.e. about 25% more time.

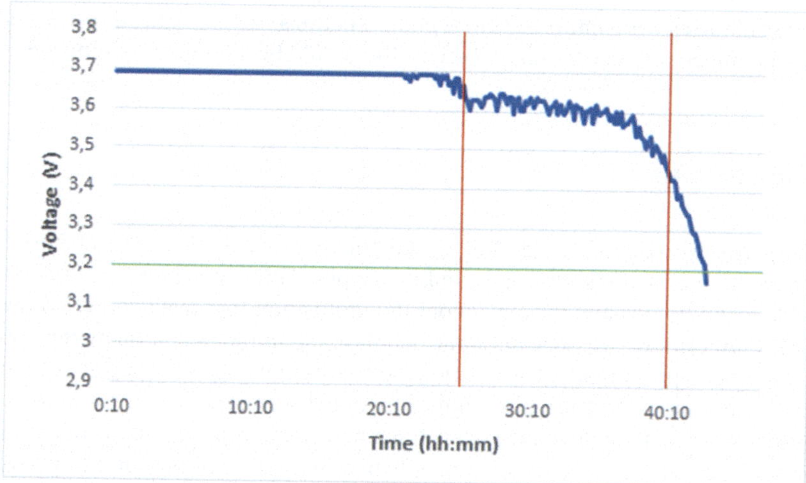

Fig. 8 Result of Test #3

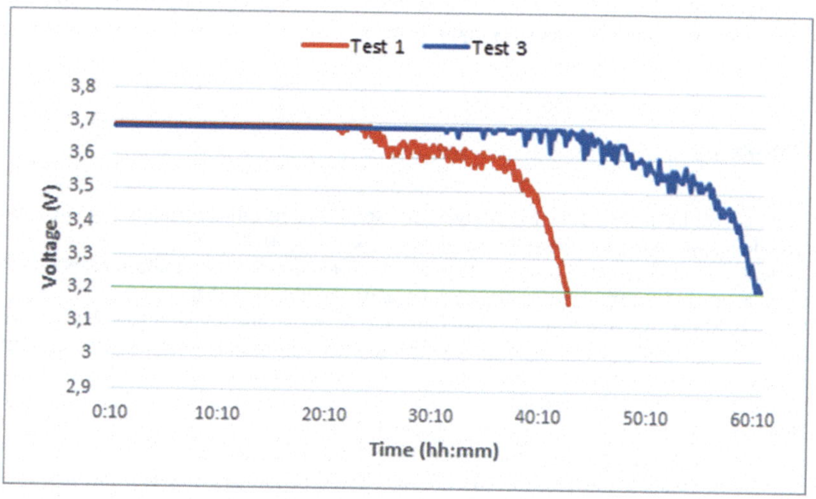

Fig. 9 Comparison

5.2.4 Comparison

The tests carried out shows that the use of encryption significantly impacts on the system's energy consumption. Comparing Test 1 and Test 3, we notice that Test 3 has a battery life 2 days longer than Test 1.

Furthermore, by comparing the two graphs (Fig. 9), we can see that the discharge process in Test 1 has a steeper curve and therefore a faster discharge. This happens because the encryption was done with functions not optimized for MCU. We expect that using libraries with optimized encryption functions (for example, those provided by ST) the impact of encryption on battery life would be significantly reduced.

6 Conclusions

In this paper we discussed the full development process of a system based on LPWANs aiming at tracking goods and/or people. The paper presents the current available technologies and reasons about the motivation that leads to the choice of NB-IoT. Furthermore, we presents some details about implementation matters of the most important apps. Finally, we reported the results of the tests aiming at evaluating power consumption and encryption impact in a real scenario.

Further work will be devoted to develop a new version of the platforms in order to increase the performance, to include a better encryption library and to performs tests on different scenarios.

Acknowledgements The authors would like to thank TIM Joint Open Lab (JOL) which made available its laboratory and the necessary equipment.

References

1. Carchiolo, V., Loria, M., Toja, M., Malgeri, M.: Real time risk monitoring in fine-art with IoT technology, pp. 151–158 (2018). https://doi.org/10.15439/2018F221
2. Carchiolo, V., Malgeri, M., Loria, M., Toja, M.: An efficient real-time architecture for collecting IoT data, pp. 1157–1166 (2017). https://doi.org/10.15439/2017F381
3. Datta, N., Malik, A., Agarwal, M., Jhunjhunwala, A.: Real time tracking and alert system for laptop through implementation of GPS, GSM, motion sensor and cloud services for antitheft purposes. In: 2019 4th International Conference on Internet of Things: Smart Innovation and Usages (IoT-SIU), pp. 1–6 (2019)
4. Eddie Chi-Wah, L.: Simple bus tracking system. J. Adv. Comput. Sci. Technol. Res. **3**(1) (2013). http://www.sign-ific-ance.co.uk/index.php/JACSTR/article/view/403
5. GSA: Narrow band IoT and M2M - NB-IoT and LTE-MTC global ecosystem and market status (2019). https://gsacom.com/paper/nb-iot-and-lte-mtc-global-ecosystem-and-market-status/. Accessed 20 April 2020
6. GSMA: Narrowband – Internet of Things (NB-IoT) (2020). https://www.telecomitalia.com/tit/it/notiziariotecnico/edizioni-2016/n-3-2016/capitolo-4.html. Accessed 20 April 2020
7. Hattarge, S., Kekre, A., Kothari, A.: LoRaWAN based GPS tracking of city-buses for smart public transport system, pp. 265–269 (2018). https://doi.org/10.1109/ICSCCC.2018.8703356
8. Huawei: GitHub – LiteOS_Lab (2020). https://github.com/LiteOS/LiteOS$_$Lab. Accessed 03 April 2020
9. IoT-Club: Narrowband – Internet of Things (NB-IoT) (2019). http://doc.iotclub.net/evbm1. Accessed 22 Mar 2019
10. Knyazev, N., Chechetkin, V., Letavin, D.: Comparative analysis of standards for low-power wide-area network. In: 2017 Systems of Signal Synchronization, Generating and Processing in Telecommunications, SINKHROINFO 2017. Institute of Electrical and Electronics Engineers Inc., United States (2017). https://doi.org/10.1109/SINKHROINFO.2017.7997528

11. Koompairojn, S., Puitrakul, C., Bangkok, T., Riyagoon, N., Ruengittinun, S.: Smart tag tracking for livestock farming. In: 2017 10th International Conference on Ubi-Media Computing and Workshops (Ubi-Media), pp. 1–4 (2017)
12. LoRa: LoRaWAN® for developers (2020). https://lora-alliance.org/lorawan-for-developers. Accessed 20 April 2020
13. Mekki, K., Bajic, E., Chaxel, F., Meyer, F.: A comparative study of LPWAN technologies for large-scale IoT deployment. ICT Express **5**(1), 1–7 (2019). https://doi.org/10.1016/j.icte.2017.12.005, http://www.sciencedirect.com/science/article/pii/S2405959517302953
14. NMEA: NMEA 0183 interface standard (2020). https://www.nmea.org/content/STANDARDS/NMEA$_$0183$_$Standard. Accessed 06 April 2020
15. Nugraha, A.T., Hayati, N., Suryanegara, M.: The experimental trial of LoRa system for tracking and monitoring patient with mental disorder. In: 2018 International Conference on Signals and Systems (ICSigSys), pp. 191–196 (2018)
16. Schacht, M.: Top 10 IoT asset tracking use cases (2018). https://www.linkedin.com/pulse/top-10-iot-asset-tracking-use-cases-maxime-schacht/. Accessed 15 April 2020
17. ShareTechnote: LTE quick reference (2020). http://www.sharetechnote.com/html/Handbook_LTE_NB_LTE.html. Accessed 20 April 2020
18. Sigfox: our story | Sigfox (2020). https://www.sigfox.com/en/sigfox-story. Accessed 20 April 2020
19. Sullings, R.: LPWAN and smart water metering (2018). https://www.linkedin.com/pulse/lpwan-smart-water-metering-rian-sullings/. Accessed 20 April 2020
20. TIM: Le tecnologie abilitanti per l'IoT (2020). https://www.telecomitalia.com/tit/it/notiziariotecnico/edizioni-2016/n-3-2016/capitolo-4.html. Accessed 20 April 2020
21. Uddin, M.S., Sumon, M., Alam, J., Islam, M.: Smart anti-theft vehicle tracking system for Bangladesh based on internet of things, pp. 624–628 (2017). https://doi.org/10.1109/ICAEE.2017.8255432

Implementing an Integrated Internet of Things System (IoT) for Hydroponic Agriculture

Georgios Georgiadis, Andreas Komninos, Andreas Koskeris, and John Garofalakis

Abstract This chapter presents ongoing work in the development of a hydroponics monitoring system by using IoT technology. Hydroponics is a method of growing plants in water based nutrient rich solution system, instead of soil. By monitoring the parameters of the solution in parallel with the environmental parameters inside the greenhouse, farmers can increase the production while decreasing the need for manual labor. Multiple networked sensors can measure these parameters and send all the necessary information to an Internet of things (IoT) platform (i.e., Thingsboard) in order the farmer or agronomist to be able to control and adjust current operating conditions (e.g. environmental controls) and plan the nutrition schedule. Furthermore Machine Learning (ML) can be used, so the system will provide recommendations to agronomists. The novelty presented in our system is that data contributed by multiple farming sites can be used to improve the quality of predictions and recommendations for all parties involved.

1 Introduction

The Internet of Things (IoT) is an active research area where sensors and smart devices facilitate the provision of information and communication. In IoT, one of the main concepts is wireless sensor networks in which data is collected from all the sensors in a network characterized by low power consumption and a wide range of

---

G. Georgiadis (✉) · A. Komninos · J. Garofalakis
University of Patras, 26504 Rio, Greece
e-mail: ggeorg@westgate.gr

A. Komninos
e-mail: akomninos@ceid.upatras.gr

J. Garofalakis
e-mail: garofalag@ceid.upatras.gr

G. Georgiadis · A. Komninos · A. Koskeris · J. Garofalakis
Diophantus Computer Technology Institute and Press, Patras, Greece
e-mail: koskeris@westgate.gr

G. Fortino et al. (eds.), *Data Science and Internet of Things*, Internet of Things,
https://doi.org/10.1007/978-3-030-67197-6_5

communication. Wireless sensor networks (WSNs) consist of multiple sensor nodes in a wireless communication-based environment. Each sensor node is to detect physical phenomena such as temperature, humidity, and moisture with limited energy and memory. WSNs are the combination of embedded systems and wireless communication which allows data transmission among the sensor nodes using radio signals. The heart of each WSN node is the microcontroller (Arduino in our case) which processes readings from its own sensors and/or readings from adjacent nodes.

IoT systems have found significant application opportunities in the agricultural sector; however, a recent review of IoT applications in agriculture has shown that research in this area is not yet fully developed [5]. For agricultural applications, IoT devices consist of sensors to measure soil properties (e.g. moisture, electrical conductivity), environmental conditions (e.g. temperature, humidity, rainfall) and ambient light conditions (e.g. solar radiation, light levels). All these parameters relate closely to the growth of plants and their monitoring, enables precision agriculture, with associated savings in energy and water consumption, as well as reduction of fertilizer and chemicals used, to support plant growth. A more recent trend in agriculture involves the move away from soil-based farming towards hydroponic agriculture. In this mode of cultivation, a greenhouse contains rows of substrate material (e.g. rockwool) on which plants are placed. Since this neutral substrate contains no nutrients, it is continuously watered with a nutrient rich solution. Hydroponic agriculture requires careful balancing of the nutrient solution contents against outdoor and indoor environmental conditions, and is thus subject to impact from even very small fluctuations away from optimal conditions. On the positive side, it is attractive to farmers because maximizes yield and provides yield level guarantees, while allowing for less use of pesticides and other control chemicals, leading to very high quality products produced in less land than would otherwise be required. On the negative side, hydroponic agriculture is far more expensive and energy-demanding compared to traditional farming [7]. The scope of this paper is to present the application of IoT systems in precision hydroponics, which is currently understudied, compared to the rest of the agricultural IoT.

2 Related Work

Hydroponic agriculture is a natural fit for IoT system applications. Setting aside backyard and small installations, professional hydroponic sites require constant monitoring and control to achieve optimal growth conditions. In typical installations, automation is provided in terms of indoor environmental controls via HVAC systems and automated windows or shading mechanisms. However, the nutrient feeding schedule is typically set by human experts (agronomists), through a feedback cycle of manual monitoring of selected plant growth (witness plants). Agronomists manually measure the growth parameters of witness plants, and adjust the feeding schedule for the whole site accordingly [10]. This, however, is a suboptimal solution. In larger sites, the microclimate can vary significantly across the site. Due to leakages, blockages and other technical issues, irrigation can be uneven. Therefore, impacts measured on witness plants, may not evenly apply to the entire site, and growth problems may go unnoticed until it is too late. In this respect, IoT can offer automated precision monitoring in real time, and across an entire site. Previous work on IoT-enabled hydroponic systems has focused mostly on system properties and architecture (e.g. [13]), with most research focusing on monitoring specific types of sensor values, such as water quality (e.g. [9]). In [15], the authors focus on the performance of the MQTT protocol, showing that server load and throughput is not significantly affected by large numbers of IoT nodes. This demonstrates that large installations can benefit from IoT deployments without needing to rely on resource-rich servers. IoT applications using very simple temperature and humidity sensors only, demonstrate that significant savings in water consumption and electricity can be obtained [11]. Similarly, in [2, 3], a system where agronomists could define logical rules for the operation of a system is presented, using a continuous feedback loop to the agronomist, so they can observe the effect of the rules they set. However, in such systems, the weak point here remains the need for heavy agronomist involvement in the process [1]. To address this problem, [16] propose a fuzzy logic control system to automatically adjust system operation, in order to maintain predetermined operational parameter specifications (in this case, solution electrical conductivity and PH). In [14], the use of machine learning is proposed in two ways, to assist hydroponic installations. Firstly, it is used to obtain an indication of witness plant growth levels by analyzing images of witness plants, replacing the need for manual measurement. Secondly, it is used to model and forecast optimal lighting policy conditions for the site, based on previous data. So far, this is the only example in literature where data has been used in an assistive manner, i.e. to automatically propose operation guidance to agronomists.

3 System Overview

For our system, we developed an embedded wireless sensor network for a hydroponic greenhouse that produces cherry tomatoes in the area of Mesolongi in Western Greece (Fig. 1). Contrary to most previous research, the system is applied to a large commercial installation. The conceptual diagram of our system is shown in Fig. 2 and consists of the wireless sensor network inside the greenhouse which transmits all the necessary environmental parameters to an Internet of things (IoT) platform.

3.1 Hardware and Sensors

The system consists of a variety of sensors that we deploy inside and outside the greenhouse and are able to measure the parameters we need, with very good precision similar to professional expensive tools. All sensors operate between 3.3 and 5.0 V.

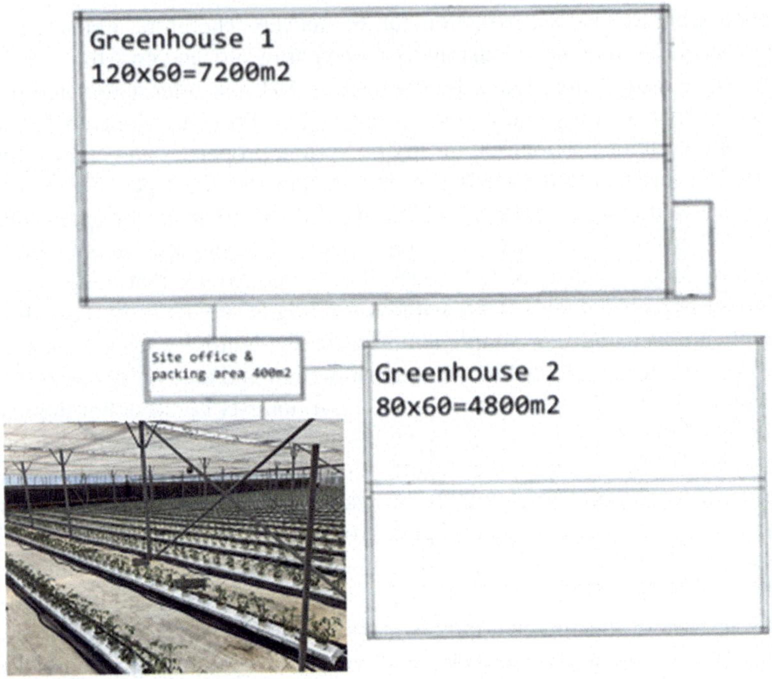

Fig. 1 The hydroponic site of our system, with a total cultivation area of 12.000 m^2 and a site office and packing area of 400 m^2

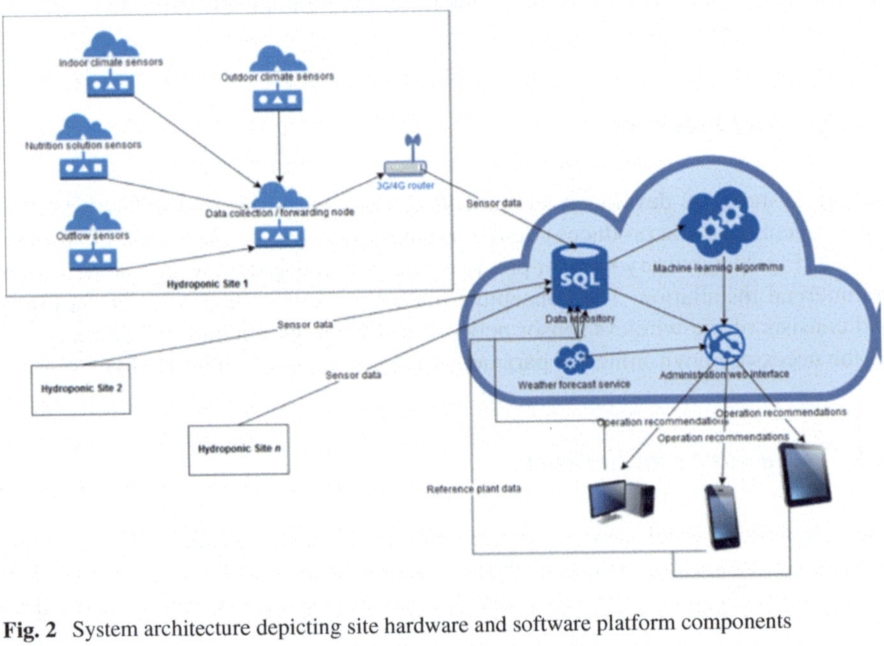

Fig. 2 System architecture depicting site hardware and software platform components

The hardware sensors were prototyped using the TinkerCAD simulator (e.g. see Fig. 3) before being implemented physically.

3.2 Indoor System Sensors

The following sensing equipment is used for indoor environmental parameter monitoring.

- **Temperature and Humidity:** A DHT-22 sensor is used for temperature and humidity levels. This sensor is a low-cost digital sensor, that uses a capacitive humidity sensor and a thermistor to measure the surrounding air. The sensor is pre-calibrated by the manufacturer and provides higher stability and accuracy compared to the DHT-11 sensor that we originally used in our system. Also, by applying appropriate open-source libraries, DHT-22 can also be used to calculate Heat Index and Dew Point of the monitoring area.
- **Ambient light and solar radiation:** Light is a basic parameter for plants growth, and needs attention regarding what and how you measure it. Most light sensors have a linear relationship with light levels, which means that they're not very sensitive to light changes. GA1A12S202 light sensor provides a log-scale relationship with light levels, compared to plain photocells, which means better response to ambient light changes. In parallel with "simple" light sensors a solar radiation sensor (SP-212 pyranometer) is used to measure in Watts per meter squared (W/m^2) total shortwave radiation from the sun. Monitoring solar radiation facilitates managing the cropping processes such as optimum time for sowing, optimum plant population, timed application of fertilizers and irrigation management [8].
- **Indoor air quality:** Indoor air quality monitoring is accomplished with CCS811 sensor that has the ability to measure eCO_2 (equivalent calculated carbon-dioxide) concentration within a range of 400–8192 parts per million (ppm), and TVOC (Total Volatile Organic Compound) concentration within a range of 0–1187 parts per billion (ppb).
- **Rockwool conditions:** For conditions inside the rockwool substrates we use two types of sensors. SEN0193 is a capacitive soil moisture sensor that can output moisture levels in a scale of 0–100% by capacitive sensing rather than resistive sensing like other sensors, and provides an overview of how "wet" is the rockwool. Although this type of sensor is mainly used in soil based plants, operates with the same accuracy in rockwool substrate. Calibration and cleaning the sensor often is needed due to the fact that contacts the different ingredients of the water based nutrition solution for long period. Also a DS18B20 waterproof temperature sensor is used for continuous monitoring the temperature inside the rockwool.
- **Power consumption:** Power consumption sensing is established by placing three Non-Invasive Current Sensors with a maximum load of 100A to each phase line (L1-2-3). SCT-013 is an ideal sensor and safe to use. By monitoring the power

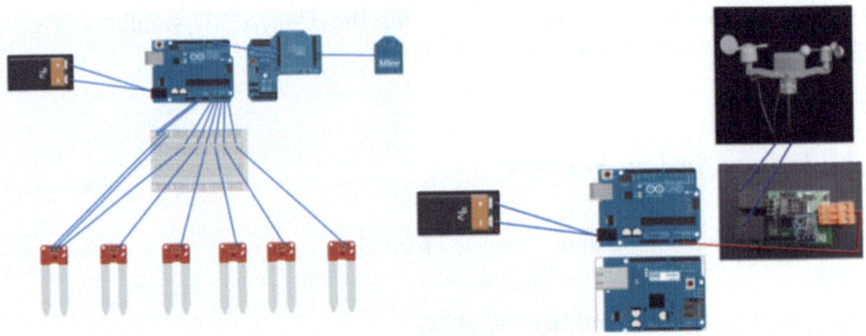

Fig. 3 Schematics for the plant row and outdoor weather station sensor packs

Fig. 4 Energy monitoring using Arduino Uno and SCT-013

consumption inside the greenhouse we can have an index of when fans, lights and irrigation are switched on (e.g see Fig. 4).

3.3 Outdoor Environment Sensors

The following sensing equipment is used for outdoor environmental parameter monitoring.

- **External environmental conditions:** For these, we use SEN0186, a mast-mounted wind speed, wind direction, and rainfall sensor with embedded temperature and moisture readings.
- **Water quality sensors:** A pH and an electric conductivity EC sensor are used on the water outflow collection points. A solution's pH measurement is a value that reflects the exact acidity or alkalinity of the solution. SEN0161 is an analogue pH meter which is commonly used in various applications such as aquaponics, aquaculture, and environmental water testing. Comes with a BNC probe that can detect a solution if it is neutral (pH = 7), acidic (pH < 7) or alkaline (pH > 7). Conductivity is the ability of a substance to carry the current and is used to measure the electrical conductivity of aqueous solutions, and then to evaluate the water quality. A DFR0300 sensor is used in our system for the measurement of water quality. From agricultural reference we know that for cherry tomatoes an ideal pH value is between 5.5 and 6.5 and for electrical conductivity around 3.7 ms/cm. Both these sensors also require calibration before first use and after long period of operation. All the above sensors are connected to different Arduino Boards and are all deployed in various specific positions inside the greenhouse.

3.4 *Virtual Sensors*

Furthermore, we support "virtual" sensors, inspired from the concept introduced in [12], where virtual sensors are described as abstractions of system components that include sampling and processing tasks. In our case, we implement such sensors through the periodic sampling and aggregation of data obtained from external APIs or other internal data sources. As a starting point we implement the following virtual sensor type:

- **Virtual weather station:** We use the OpenWeathermap.org API to obtain current and forecasted meteorological conditions in the area. This data is used for interpolation with outdoor sensor data, and to obtain weather forecasts from established climate models.
- **Nutrition scheduling:** The agronomist maintains an accurate record of the nutrition schedule, including watering times and duration, and nutritional content composition, in the form of spreadsheet files. These data are fed to the system at regular intervals.
- **Manual pH, EC and WC readings:** The agronomist takes manual pH, electric conductivity and water content readings twice daily. We store this data in a virtual sensor device, through use of a dedicate mobile app, at the time of measurement.
- **Witness plant growth:** The agronomist takes daily measurements of specific plants (e.g. number of flowers, number of tomatoes, plant length etc.). We store this data in a virtual sensor device, through use of a dedicate mobile app, at the time of measurement.

3.5 Sensor Integration and Connectivity

We bundle multiple sensors on Arduino Uno boards, building what we call "sensor packs" with various sensor configurations. These various sensor packs connect via XBee S2C RF modules that are placed on suitable XBee shields in order to form a mesh network configuration. The XBee formed network consists of sensing nodes and coordinator nodes that aggregate and forward data to the server, via wired Ethernet connections [6]. The following wireless sensor pack configurations are supported (e.g. Figs. 3 left and 5).

- **Plant row packs:** These support multiple (6) soil moisture sensors and, optionally, a soil temperature sensor. These sensors are positioned along individual plant rows. Indoor climate packs: These support the temperature, humidity, air quality, light level and solar radiation sensor. They are placed in various equidistant locations across the site.
- **Outflow packs:** These support the pH and electric conductivity sensor and are placed in the outflow collection points across the entire site.

We also support the following wired Ethernet pack configurations (e.g. Fig. 3 right).

- **Weather station pack:** One Arduino board is used to integrate all the outdoor sensors for the site. This sends data directly to the server using a wired Ethernet connection.
- **Energy monitor:** A three-phase energy meter Arduino is placed inside the main power supply board and monitors the power loads in each phase.
- **Coordinator packs:** Each coordinator node collects data wirelessly (via XBee) from multiple other sensor packs (plant row, indoor climate and outflow. Its role is to collect, store, pre-process and transmit the data to the server, using a wired Ethernet connection. Because of the heavier computation demand, coordinator packs are integrated using the Arduino Mega board.

3.6 Manual Data Collection

The greenhouse agronomist takes regular observations from the various witness plants in the site, and using manual pH, EC and WC instruments. These observations are collected using a mobile device (tablet), running an application to collect the relevant data (foliage, plant height, stem width, fruit size and state etc.). For this part we developed a basic "Greenhouse app" that can be installed at the moment only for Android devices and allows the user/agronomist to insert his observations on a daily basis for specific plants that are monitored. Users can add a set of witness plants manually by selecting first the position where the plant was originally placed and then to add parameters or attributes to a form for observation (e.g. Fig. 6). Data is uploaded directly to the server using a wireless (Wi-Fi) connection.

Fig. 5 Deployed plant row sensor pack

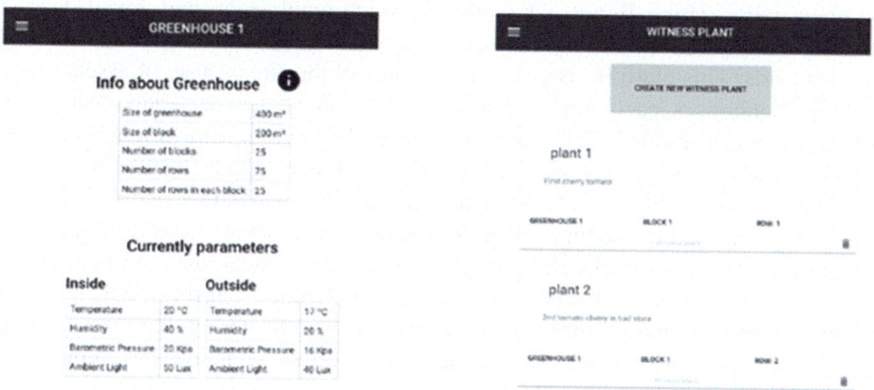

Fig. 6 Information screen for GreenHouse 1 (left) and witness plant screen (right)

The following diagram represents the steps the user can follow when using the app and all the possible functionalities. After first sign in, there is a menu with the options sensors, greenhouses and witness plant, with each option to have its own sub-options (e.g. Fig. 7).

3.7 *IoT Data Management Platform*

Currently, there are no global IoT standards for the description of data and communication across the various IoT system layers (sensors, middleware, back-end) [4]. Each system is responsible for its own interoperability standards, and one of the most

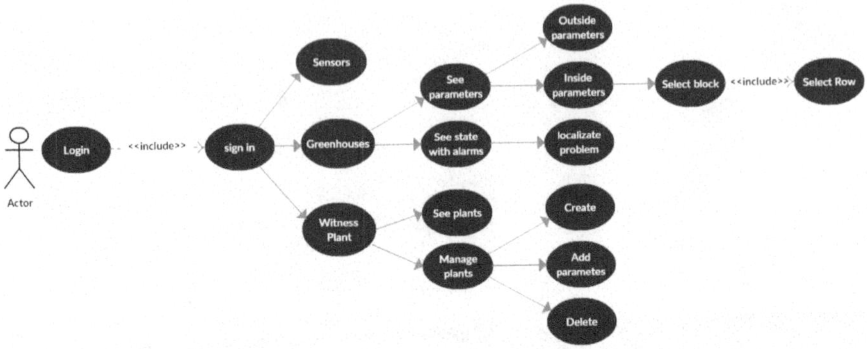

Fig. 7 Tree diagram of greenhouse app

common techniques is to comply with standards set by the IoT data management platform selected by the developers. Across the various options, we selected to use the open-source ThingsBoard IoT platform, which revolves around a well-defined data API standard for communication and data exchange, based on JSON-structured data payloads. This platform is configured to model the site as a set of assets (there are two greenhouses, each one being an individual asset), devices (each sensor pack or virtual sensor is modelled as a single device) and operators with various roles (site supervisor, agronomist). The platform offers the ability to visualise the data at an aggregate or individual sensor basis (using custom-designed data visualisation dashboards, Fig. 13), and to establish alerts for operating conditions exceeding specified thresholds. The sensor packs and virtual sensors communicate data to the platform using HTTP REST APIs. Further from the data collection platform, ML modules for training models and performing predictions are also hosted on the server. These modules are currently under development, but the aim is to feed their output back into the platform, and integrate their output on the user's dashboard, in numeric and graphical form.

3.8 *System Integration Validation*

Implementation of our system started in early November 2018 in parallel with the planting of the new crop. It was considered necessary to start recording the parameters that are directly related to the growth of plants in the greenhouse. The cultivation of the specific type of tomato is almost annual, which gave us useful information regarding the plants and the environmental conditions inside the greenhouse throughout the year. In late August 2019 the plants were removed, the greenhouse was cleaned and the new plants were transferred on the new rockwool by late September. This was a period that gave us the opportunity to clean and re-calibrate most of the sensors or even to adjust the topology of our sensor network. In the following sections, we outline some of the practical issues that we encountered during the installation and integration tests of our system.

3.8.1 Sensor Board Integration Issues

As described previously, our system is based on commercial off-the-shelf components, which we had to integrate into a single system. While theoretically this approach should be trouble-free, in practice we encountered several issues that mostly concern the hardware components. One of the main issues we encountered in the beginning of our implementation was the use of Arduino-compatible Ethernet Shields using the W5100 chip. We use this Ethernet Shield in XBee coordinators, the external weather station and also in Energy monitoring sensor packs. The shield allows transfer of data to our system database. This type of shield comes unconfigured, so the user has to specify a unique MAC ADDRESS inside the code for network connectivity. After two or three days of normal operation, we observed that data was not being sent to our database. A closer inspection indicated that the Network Router couldn't assign static or dynamic ip addresses to these Ethernet shields anymore. Even when we replaced the Network Router we came across the same problem after a few days. The solution to this network issue was to replace all Ethernet Shields with other versions that use the W5500 chip. These shields come with already assigned MAC addresses and can operate without problem continuously.

3.8.2 Heat-Induced Sensor Operation Issues

All Arduino based boards that are placed inside the greenhouse are inside plastic enclosures in order to avoid water contact and humidity. The plastic enclosures are not totally waterproofed because of small holes that are drilled for the purpose of air circulation inside the case. During the summer period and more specific between July and August, due to high external temperatures (averaging 36 °C around 3pm), the temperature inside the greenhouse can reach a value of 38 °C , which led three of the indoor environment sensor packs that were placed above the curtains to "freeze" due to high temperature. As the time progressed towards the afternoon, and the position of the sun was not directly above the greenhouse, the temperature inside the greenhouse dropped to around 30 °C, these specific Arduino-based sensor packs reverted back to normal operation by themselves. A solution to this problem for the summer period was to adjust the position of the sensor packs below the curtain of the greenhouse in order to afford them additional protection from sunlight-induced high temperatures, as well as to drill additional small holes through their plastic enclosures for more air circulation.

(a) Plant height at start of planting season (b) Plant height at 6 months after planting

Fig. 8 Plant growth interferes with wireless sensor communication

3.8.3 Operation Problems Due to Plant Growth

As we mentioned before implementation of our system started when the new crop was planted. At this time the plants have an average of 25 cm in length (Fig. 8a). At this point there is enough space for placing all the necessary hardware (Arduino boards, sensors and network cables) that allow data collection, and more important wireless communication between nodes of the network. As the plants grow, we have to observe their size expansion which is critical for our hardware operation. As we can see in Fig. 8b, after six months the plants have gain an average length of 1.5 m, which leads to aerial signal failure (LoS) between boards, and most of the times requires replacing our boards to new positions inside the greenhouse in order to achieve communication. Also due to root development sometimes the boards are trapped inside the roots which make it difficult to access them without breaking the root at some point (Fig. 9).

3.8.4 Humidity Effects

Apart from high temperatures that they were recorded during the summer period, high level of humidity also was an issue regarding our hardware equipment. Especially around the spring period and more specific at around 8a.m, humidity levels inside the greenhouse reached about 85%. This was noticed in a form of water drops and pools on the top of the (supposedly waterproof) enclosures where we keep all network and power supply equipment (Fig. 10a). Humidity also tends to form inside the enclosures, affecting the rubber seal adhesive, and weakening the ability to keeping water and humidity outside (Fig. 10b). In order to overcome this issue, we had to apply a layer of silicone to all joints of our equipment boxes to keep humidity outside.

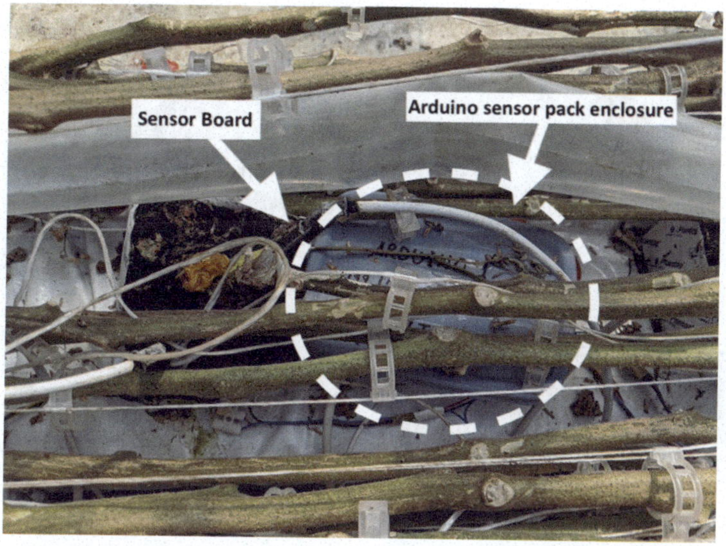

Fig. 9 Root growth interfering with system wiring and access to sensor boards

(a) Water condensation on the network equipment enclosure

(b) Humidity effects on internal enclosure insulation strips

Fig. 10 Humidity problems during greenhouse operation

Fig. 11 Manual workers during normal greenhouse operations

3.8.5 Labourer Interference with System

Every day the greenhouse follows a scheduled timetable of operation. Watering the plants, collecting tomatoes that are ready or even adjusting the height of the plants in order to grow in a way that can give the maximum height (Fig. 11). During this process, we found that often manual labourers would damage a sensor, or relocate a sensor to a point that it couldn't measure the desired value anymore. A solution to this problem was to relocate the sensor packs to a position high enough inside the greenhouse in order not to disturb the plant growth and also interfere less with the workers daily routines.

3.8.6 Sensor Measurement Quality Due to Rockwool Conditions

In hydroponics, rockwool is the substrate where the plants are placed and grow. This substrate during the summer period can be watered about forty times a day. It has the ability to absorb water and allow the roots to grow inside. As we mentioned in 3.2, to monitor rockwool conditions we use SEN0193 sensor. After calibrating the sensor, we place the sensor inside the rockwool and starts to send moisture levels to our database. A normal accepted value is between 75 and 95%. After a short period usually three to five days, the values could drop to 25%. This happens due to the expansion of the rockwool which leads the sensor to become loose and not to be in touch with the substrate. These minimum values of moisture are an alarm to our system to relocate the sensor. Furthermore, due to the fact that the sensor most of the time is continuously in touch with water, corrosion is also an issue that leads to abnormal moisture values (Fig. 12).

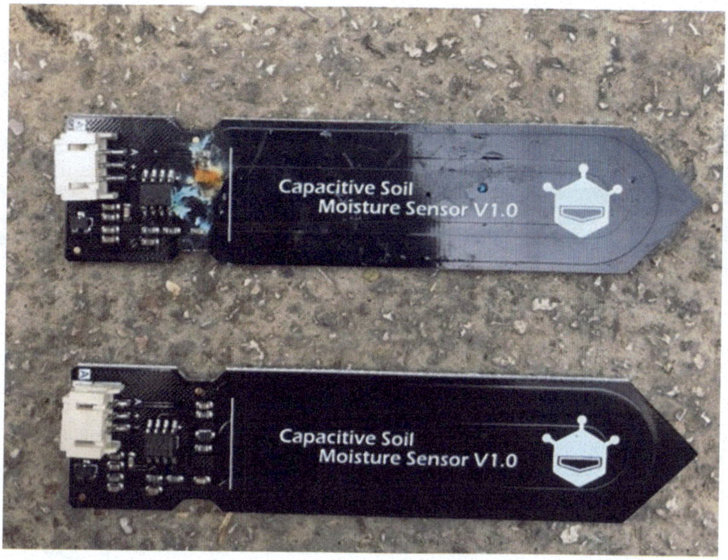

Fig. 12 Corrosion on soil moisture sensor after prolonged use (top) versus new sensor (bottom)

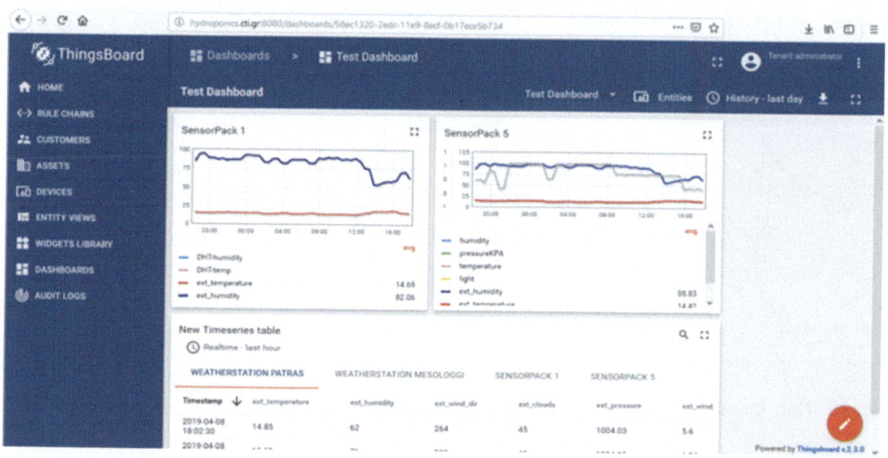

Fig. 13 Sample view of the hydroponic sensor dashboard provided by the ThingsBoard IoT platform

4 Monitoring Actuator Operation

The hydroponic site is equipped with a proprietary climate control system (by Priva S.A.) which is used to automatically drive various control elements that help maintain optimal indoor climate parameters, as required. These include ventilation fans, over-

(a) Under curtain positioning (b) Over curtain positioning

Fig. 14 Positioning of indoor environment sensor packs to detect system operation

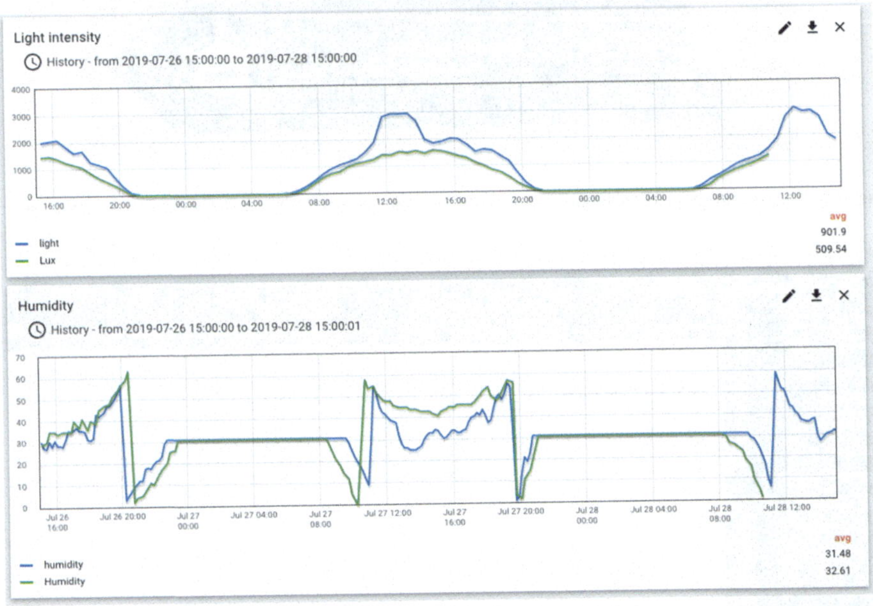

Fig. 15 Data captured by the over-curtain and under-curtain indoor environment sensor packs

head curtains, moisture sprayers and CO2 pumps, which are activated depending on specific rules set by the agronomist. This proprietary system does not have an open interface, therefore it is not possible to access data produced by it from a third-party system such as ours. However, strategic positioning of sensors can be used to infer the system operation and to provide an overview of the system state and function, visualised through a control dashboard (Fig. 13).

As an example, we demonstrate the data acquired by two indoor environment sensor packs, positioned over, and under the overhead curtains (Fig. 14). In this example, the green line shows the under-curtain sensor values, while the blue line shows the over-curtain values. As can be seen in Fig. 15, when the solar radiation intensity spikes

Table 1 Deep learning model evaluation

Evaluation approach	Root mean squared error	Correlation (Spearman's ρ)
k-fold cross validation	$0.114(\sigma = 0.091)$	$0.877(p < 0.01)$
Leave-one-out	$0.057(\sigma = 0.086)$	$0.892(p < 0.01)$

(blue line, top chart), the automatic curtains begin to operate, therefore "smoothing" the effect of intense light in the greenhouse. Simultaneously, because of the high light intensity and resulting increased temperature, humidity begins to drop rapidly, hence the moisture sprayers help maintain the humidity levels within better tolerances. By capturing this behaviour, we are able to detect system operation events, as well as threatening emergent conditions in the greenhouse during operation.

5 Machine Learning Experiment

Our work regarding the integration of ML recommendation components is ongoing. However, in this section, we demonstrate an example of how ML can provide advanced insights and recommendations to hydroponic site agronomists, based on the prediction of ambient light levels in the greenhouse. We use a deep neural network model on a six day dataset, ranging between 27/4/2019 and 2/5/2019 (144 cases), with the layer node count $\{3, 5, 5, 5, 3\}$ ReLU activation function and $\varepsilon = 1.0^{-8}$ and $\rho = 0.99$. Input features are the hour of day ([0, 23]), and reported cloud coverage level ($[0\%, 100\%]$) from the virtual weather station, while the predicted value is the ambient light levels in the greenhouse in lux units ($[0-, +\infty)$). Raw data is aggregated to obtain their hourly average. Data preprocessing includes the normalisation of lux values to a range between [0, 1]. The deep learning model can produce arbitrary positive or negative values. The latter, in our case, make no logical sense (since lux values cannot be below 0). Hence we post-process the predictions to transform negative values to zero. To evaluate the model performance, we use k-fold cross validation ($k = 10$) withrandom sampling, and also use a leave-one-out approach. The results are shown in Table 1 and Fig. 16. Interestingly, just two features (cloud cover, hour of day) are reasonable predictors for light intensity (between 5.7 and 11.4% RMSE), despite inaccuracies caused by plant foliage, worker and equipment movement and the OpenWeatherMap API model inaccuracy. Further, lux data in this analysis comes from a single, uncalibrated sensor, whose values are not cross-related with those of nearby sensors.

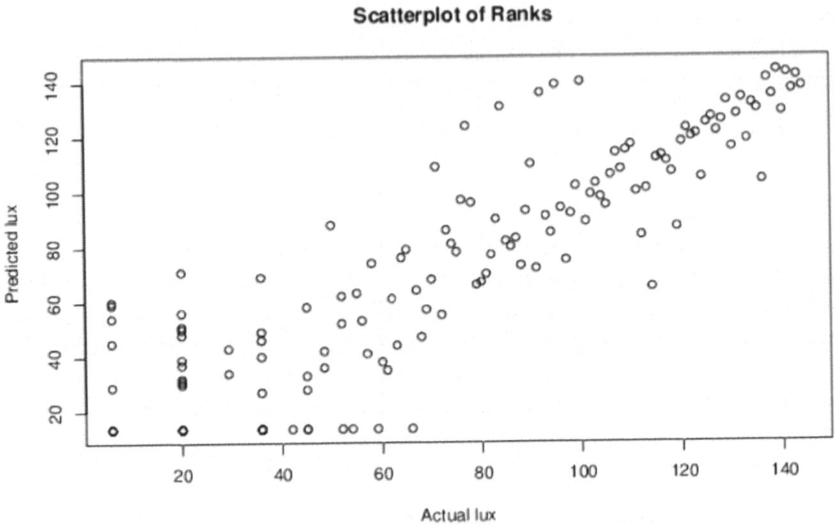

Fig. 16 k-fold cross validation Spearman rank correlation

6 Conclusion

We have presented the architecture and on-going deployment effort for an IoT-enabled hydroponic installation. Our work aims to be amongst the first to present a complete system that can produce recommendations to facilitate the workload of professional agronomists, using ML techniques.

Still there are a few issues that need to be overcome. A clean output data from the sensors is mandatory for our system, and most of the sensors require frequent cleaning and calibration for achieving this. Also plant growth affects the communication between sensor nodes (LoS) which leads to repositioning the nodes in new places keeping in mind not to interfere with the greenhouse's everyday program, and this action most of the times demands reconfiguration of the sensor network.

A further innovation which we are in the process of implementing is to enhance the quality of predictions by collecting data from multiple contributing sites. In this way, a community of hydroponic installation operators can benefit from the knowledge contributed by others, and therefore solving the "starting problem" associated with the inability to obtain recommendations when no, or little data, has been obtained for the site. A further advantage of this approach is that it may minimize the need for IoT equipment installation, as fewer sensors on each site will be necessary to obtain accurate monitoring and prediction results.

As the project draws to an end, we plan to release a dataset from a full cultivation period to the wider research community. We hope that this shall help future researchers with training their own models for cultivation predictions, anomalous event detection and other related purposes.

Acknowledgements This project was co-financed by the European Regional Development Fund and the Government of Greece through the NSRF2014-20 program (Western Greece RIS3 Microelectronics and Advanced Materials, project code ΔΕ1Μπ-0060). The authors would also like to thank Nikos Kanakaris and Georgios Kournetas for contributing in the composition of the LaTeX code of this paper.

References

1. Crisnapati, P.N., Wardana, I.N.K., Aryanto, I.K.A.A., Hermawan, A.: Hommons: hydroponic management and monitoring system for an IOT based NFT farm using web technology. In: 2017 5th International Conference on Cyber and IT Service Management (CITSM), pp. 1–6 (2017). https://doi.org/10.1109/CITSM.2017.8089268
2. Ferrández-Pastor, F.J., García-Chamizo, J.M., Nieto-Hidalgo, M., Mora-Pascual, J., Mora-Martínez, J.: Developing ubiquitous sensor network platform using Internet of Things: application in precision agriculture. Sensors **16**(7), 1141 (2016). https://doi.org/10.3390/s16071141
3. Ferrández-Pastor, F.J., García-Chamizo, J.M., Nieto-Hidalgo, M., Mora-Martínez, J.: Precision agriculture design method using a distributed computing architecture on Internet of Things context. Sensors **18**(6), 1731 (2018). https://doi.org/10.3390/s18061731
4. Fortino, G., Savaglio, C., Palau, C.E., Suarez de Puga, J., Ganzha, M., Paprzycki, M., Montesinos, M., Liotta, A., Llop, M.: Towards multi-layer interoperability of heterogeneous IoT platforms: the INTER-IoT approach. Integration, Interconnection, and Interoperability of IoT Systems. Springer International Publishing (2018). https://www.springerprofessional.de/en/towards-multi-layer-interoperability-of-heterogeneous-iot-platfo/13304824
5. Gómez-Chabla, R., Real-Avilés, K., Morán, C., Grijalva, P., Recalde, T.: IoT applications in agriculture: a systematic literature review. In: Valencia-García, R., Alcaraz-Mármol, G., del Cioppo-Morstadt, J., Vera-Lucio, N., Bucaram-Leverone, M. (eds.) ICT for Agriculture and Environment. Advances in Intelligent Systems and Computing, pp. 68–76. Springer International Publishing, New York (2019)
6. Kooijman, M.: Building Wireless Sensor Networks Using Arduino. Packt Publishing, Birmingham (2015). https://www.oreilly.com/library/view/building-wireless-sensor/9781784395582/
7. Lages Barbosa, G., Almeida Gadelha, F.D., Kublik, N., Proctor, A., Reichelm, L., Weissinger, E., Wohlleb, G.M., Halden, R.U.: Comparison of land, water, and energy requirements of lettuce grown using hydroponic vs. conventional agricultural methods. Int. J. Environ. Res. Public Health **12**(6), 6879–6891 (2015). https://doi.org/10.3390/ijerph120606879, https://www.ncbi.nlm.nih.gov/pmc/articles/PMC4483736/
8. Maldonado-Guzmán, G.: Intelligent Agriculture: Developing a System for Monitoring and Controlling Production. Emerald Publishing Limited, Bingley (2019). https://books.emeraldinsight.com/page/detail/Intelligent-Agriculture/?k=9781789738469
9. Manju, M., Karthik, V., Hariharan, S., Sreekar, B.: Real time monitoring of the environmental parameters of an aquaponic system based on Internet of Things. In: 2017 Third International Conference on Science Technology Engineering Management (ICONSTEM), pp. 943–948 (2017). https://doi.org/10.1109/ICONSTEM.2017.8261342
10. Pantanella, E.: Aquaponics production, practices and opportunities. In: Hai, F.I., Visvanathan, C., Boopathy, R. (eds.) Sustainable Aquaculture, Applied Environmental Science and Engineering for a Sustainable Future, pp. 191–248. Springer International Publishing, Cham (2018)
11. Pitakphongmetha, J., Boonnam, N., Wongkoon, S., Horanont, T., Somkiadcharoen, D., Prapakornpilai, J.: Internet of things for planting in smart farm hydroponics style. In: 2016 International Computer Science and Engineering Conference (ICSEC), pp. 1–5 (2016). https://doi.org/10.1109/ICSEC.2016.7859872

12. Raveendranathan, N., Galzarano, S., Loseu, V., Gravina, R., Giannantonio, R., Sgroi, M., Jafari, R., Fortino, G.: From modeling to implementation of virtual sensors in body sensor networks. IEEE Sens. J. **12**(3), 583–593 (2012). https://doi.org/10.1109/JSEN.2011.2121059
13. Ruengittinun, S., Phongsamsuan, S., Sureeratanakorn, P.: Applied internet of thing for smart hydroponic farming ecosystem (HFE). In: 2017 10th International Conference on Ubi-Media Computing and Workshops (Ubi-Media), pp. 1–4 (2017). https://doi.org/10.1109/UMEDIA.2017.8074148
14. Somov, A., Shadrin, D., Fastovets, I., Nikitin, A., Matveev, S., Seledets, I., Hrinchuk, O.: Pervasive agriculture: IoT-enabled greenhouse for plant growth control. IEEE Pervasive Comput. **17**(4), 65–75 (2018). https://doi.org/10.1109/MPRV.2018.2873849
15. Triawan, M.A., Hindersah, H., Yolanda, D., Hadiatna, F.: Internet of things using publish and subscribe method cloud-based application to NFT-based hydroponic system. In: 2016 6th International Conference on System Engineering and Technology (ICSET), pp. 98–104 (2016). https://doi.org/10.1109/ICSEngT.2016.7849631
16. Yolanda, D., Hindersah, H., Hadiatna, F., Triawan, M.A.: Implementation of real-time fuzzy logic control for NFT-based hydroponic system on Internet of Things environment. In: 2016 6th International Conference on System Engineering and Technology (ICSET), pp. 153–159 (2016). https://doi.org/10.1109/ICSEngT.2016.7849641

A Collaborative BSN-Enabled Architecture for Multi-user Activity Recognition

Qimeng Li, Raffaele Gravina, Congcong Ma, Weilin Zang, Ye Li, and Giancarlo Fortino

Abstract Human activity plays a significant role in various fields, such as manufacturing, healthcare, and public safety; therefore, recognizing human activity is crucial to enable smart innovative services. The development of ubiquitous sensing and pervasive computing allows studying what humans perform in real time and mobility. Single- and multi-user activity recognition (AR) differ by the number of involved users. With recent developments of multi-sensor and multi-information fusion, multi-user activity recognition is gradually becoming an emerging and relevant research frontier. In this paper, we propose a software architecture which combines cloud and edge computing with collaborative body sensor networks (CBSNs) to support the development of CBSNs-enabled services and in particular we provide its case-study in the context of multi-user AR.

Q. Li (✉) · R. Gravina · G. Fortino
Department of Informatics, Modeling, Electronics and Systems, University of Calabria, 87036 Rende, Italy
e-mail: qimeng.li@unical.it

R. Gravina
e-mail: r.gravina@dimes.unical.it

G. Fortino
e-mail: fortino.giancarlo@unical.it

C. Ma
Nanyang Institute of Technology, Nanyang, China
e-mail: macc@whut.edu.cn

Q. Li · W. Zang · Y. Li
Shenzhen Institutes of Advanced Technology, Chinese Academy of Sciences, Shenzhen, China
e-mail: wl.zang@siat.ac.cn

Y. Li
e-mail: ye.li@siat.ac.cn

G. Fortino et al. (eds.), *Data Science and Internet of Things*, Internet of Things,
https://doi.org/10.1007/978-3-030-67197-6_6

1 Introduction

In the last decades, with the continuous development of microelectronics and the increasingly diffused Internet of Things (IoT) technology, more and more sensors are being pervasively spread in daily living environments. Estimates provided by IHS (Information Handling Services) predict that " The IoT market is forecast to grow from an installed base of 15.4 billion devices in 2015 to 30.7 billion devices in 2020 and 75.4 billion in 2025 [1]". With these smart objects, a wide range of information can be conveniently acquired. Such information can be used in many domains, including public safety, industrial manufacturing, and healthcare.

In this emerging technological shift, the concept of human-centric services will increasingly appear in daily life and workplace. Furthermore, people conceptually and physically interact and influence each other to produce various behavior patterns [2]. Human activity, recognized by these ubiquitous sensors, therefore, is critical in all of the areas where humans are the core actors. However, literature of human activity recognition (HAR) [3–5] is focused on individual activities [6–10], while multi-user activities [11] are scarcely studied. HAR becomes extremely complicated in multi-user scenarios although precious high-level information could be extracted. For more than a decade, the problem of recognizing human actions and activities using cameras [12–14] has been studied in computer vision [15–17], with controversial results, especially in unconstrained real-world environments. Besides, sensors show significant potential to support HAR by capturing useful low-level features, such as human motion, physiological signals, environmental parameters, etc. However, recognizing human activities using sensors is challenging because sensor data are inherently noisy and human activities are often a complex composition of individual movements. Recent studies are mostly focused on single-user AR (frequently addressed with wearable sensors). On the contrary, research on multi-user AR is in embryonic stage yet. Multi-user AR is a non-trivial extension of single-user AR since it requires the joint analysis of two or more individuals in a shared environment that may interact and collaborate performing actions jointly. Specifically, multi-user activity can be divided into three scenarios: the first considers multiple users (in the same environment) performing activities individually (e.g. having lunch together); the second involves a group interacting and physically collaborating to achieve a shared goal (e.g. lifting a heavy weight from the floor); the third is a hybrid scenario which mixes individual and group activities.

In this paper, we aim at more effectively supporting the development of collaborative multi-user AR, with the design of a novel BSN-enabled architecture based on BodyCloud [18], BodyEdge [19], and the concept of CBSNs [20] to achieve:

- distance and communication latency reduction between users and servers and higher scalability, by integrating cloud and edge infrastructure;
- increased responsiveness, through cloud and edge computing resources;
- co-located collaboration and interaction recognition.

The remainder of the paper is organized as follows. Section 2 discusses the related work on recent multi-user AR. Section 3 describes the proposed multi-user AR architecture. Section 4 reveals the building blocks of the programming model for multi-user AR. Finally, Sect. 5 concludes the paper and outlines planned future work.

2 Related Work

Most of the current related literature focuses on identifying single-user activities. However, in real life, humans often form temporary and persistent groups to achieve specific tasks. The lack of a systematic approach of multi-user AR could be effectively addressed thanks to the diffusion of IoT technologies—and the edge paradigm in particular—since environment/context awareness plays a crucial role in multi-user scenarios. An effective multi-user AR method can be successfully exploited in diverse application domains including public safety, industrial manufacturing, and health care.

So far, researchers used computer vision-based approaches to recognize multi user activities. In [12], the authors used a discriminative temporal interaction manifold framework to characterize the group motion pattern, and experiments on football play recognition demonstrate the effectiveness of the approach. Ibrahim et al. [13] proposed a deep learning model to capture group AR based on long short-term memory (LSTM) models. Deng et al. [14] combined graphical models with deep neural networks; experimental results on group AR demonstrate the potential of the model to handle highly structured learning tasks. However, computer vision methods are still raising privacy issues. Besides, few studies addressed human-centric recognition using wireless body sensors. Wang et al. [21] developed a wearable sensor platform to collect sensor data for multiple users; two temporal probabilistic models (i.e. Coupled Hidden Markov Model (CHMM) and Factorial Conditional Random Field (FCRF)) were used to classify the multi-users activity. Alhamoud et al. [22] utilized a technique of multi-label classification to recognize activity in multi-user environment; data were acquired from appliance-level power sensors as well as environmental sensors in a two-person apartment. Augimeri et al. [20] developed the E-Shake system using Shimmer [23] heart rate and motion sensor to recognize handshaking between two people and the mutual emotional reactions while shaking hands. Prossegger et al. [24] adopted an approach based on decision trees (E-ID5R) to recognize multi-resident activity. In [25], authors proposed a distributed middleware called GroupSense to recognize several group physical activities which obtain knowledge from individual activities and movements of smart objects (IoT devices).

Thanks to their small size, pervasive application, and convenient cost, wearable sensor devices have become the most common solution to HAR. However, since multi-modal sensors are capable of generating significant amount of data in real time streams, it is not possible to exclusively rely on their computing resources, especially for multi-user HAR where the complexity of the problem remarkably raises. To tackle this issue, Cloud and edge computing represent a solid and effective solution [18, 19].

Therefore, the proposed approach combines cloud and edge computing resources to process large amounts of data; it reduces communication latency and provides wide scalability with quick responsiveness. While the proposed architecture is currently applied for multi-user AR recognition, its general-purpose design principles allow it to be effective in other application collaborative domains, such as contextual interaction, affective and cognitive computing.

3 An Architecture for Multi-user Activity Recognition

The proposed human-centric CBSN-enabled architecture, whose high-level schema is depicted in Fig. 1, provides useful features and information by implementing distributed CBSNs and exploiting smart environments, edge computing, and cloud services. Such systems are characterized by the combined use of software components with heterogeneous physical devices and protocols, so as to recognize multi-users activities and/or to provide final users with high-level services.

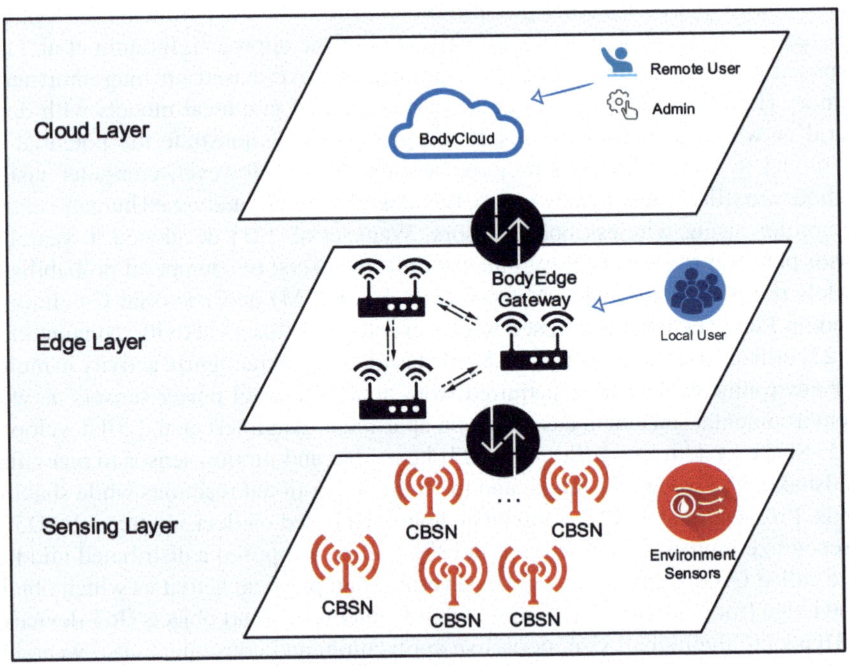

Fig. 1 A hierarchical structure for recognizing multi-user activities

3.1 BodyCloud

The BodyCloud [18] is an open-source Cloud-assisted middleware for pervasive and continuous monitoring of assisted livings using wearable and mobile devices.

In our design, we selected the BodyCloud as an optional component at Cloud layer, which comes into play essentially in just two cases:

1. when the user needs long-term monitoring and historic and statistical analysis;
2. when there are external users and administrators who need to access data or control the system remotely.

3.2 BodyEdge Gateway

Edge computing paradigm moves part of application logic, data and computing power (services) from the core to the boundary of the network. To a certain extent, it reduced transmission cost and network delay between users and servers. BodyEdge [19] is a specific platform used to recognize multi-user activity.

The BodyEdge Gateway (BEG) (see Fig. 2) is used in the proposed architecture as an intermediate layer to connect CBSNs and BodyCloud together. The functional scheme of the BEG is constituted by the following modules:

- *Edge Access Interface:* it allows the communication between BEG and CBSNs for exchanging data and sending control commands.
- *CBSN Directory:* it maps the names of CBSN resources to their respective addresses.
- *CBSN Management:* it is in charge of periodically executing a discovery procedure to find new CBSNs asking for a connection to the BEG and to register CBSNs with their devices in the BEG.
- *Storage Service:* when enabled, it stores all the data received from CBSNs and generated by BEG.
- *BodyEdge Gateway Manager:* it handles local data which does not need to be uploaded to Cloud resources or facilities. High-level data processing or mining functionalities such as data fusion or classification will be used for real-time local processing and computing.
- *Remote Communication Interface:* it allows the communication between BodyCloud and BEG for exchanging data and to send/receive control commands from the upper Cloud layer.

3.3 Collaborative Body Sensor Networks

CBSNs [5] are specific BSNs [26–28] with cooperative behavior providing co-located users with collaborative and distributed application services. The CBSN

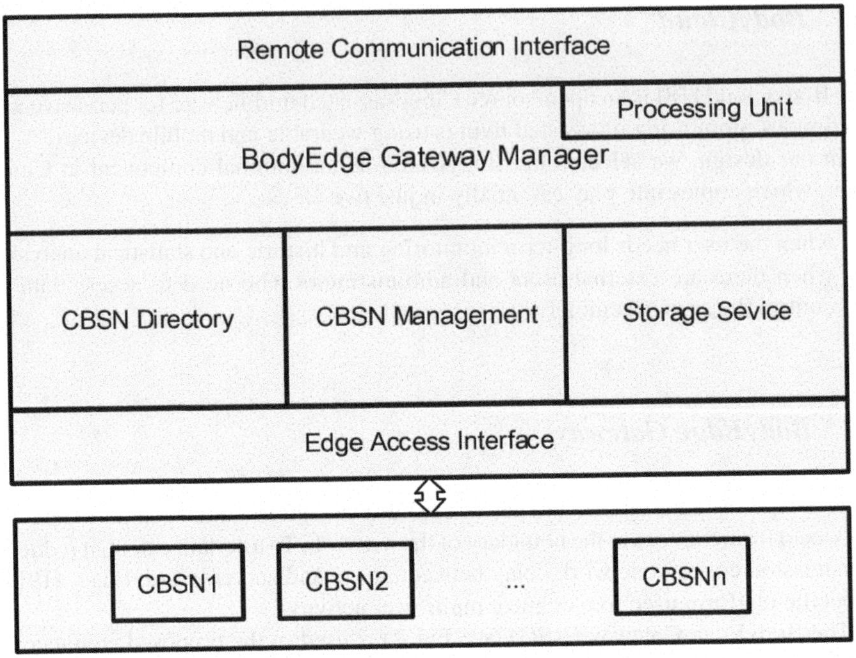

Fig. 2 Architecture of BodyEdge gateway

reference architecture is shown in Fig. 3. In particular, a CBSN is composed by a base station (BS) (also called personal coordinator) which manages a set of wearable sensors (WSs) through a single-hop communication protocol (e.g. Bluetooth, ANT+, Zigbee).

The presented CBSN consists of the following components:

- *CBSN Manager:* it decides which data needs to be uploaded to BEG and handles local data which does not need to be uploaded to BEG. Low-level data processing such as feature extraction will be used for real-time local processing and computing.
- *Collaboration Services:* it enables collaborative computing among CBSNs to implement collaborative applications.
- *Edge API:* it allows the communication between the CBSNs and BEG for exchanging data and to send/receive control commands.
- *Peer-to-Peer Communication:* it allows the direct communication among CBSNs for exchanging data.
- *Service Selection and Actuation:* it combines semi-automatic procedures (partly depend on user intervention) and rules to select and activate services among neighbor CBSNs.
- *CBSN Presence and Discovery:* it allows to detect neighbor CBSNs efficiently and enables the discovery services to execute.

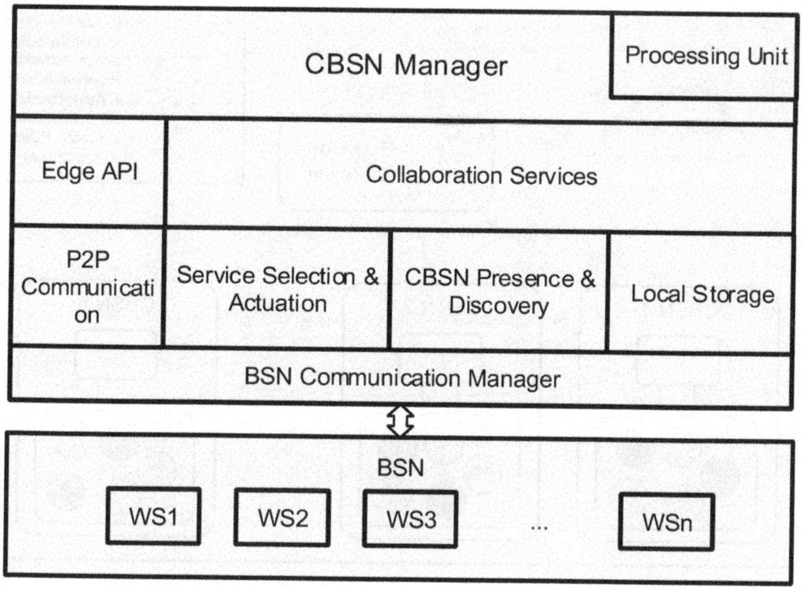

Fig. 3 Architecture of CBSN

- *Local Storage:* when enabled, it stores all the data received from WS nodes and generated by CBSN.
- *BSN Communication Manager:* it allows the communication between coordinator (BS) and CBSNs for exchange data and sending control commands.
- *BSN:* self-organized body area network, typically composed of one or multiple wearable devices.

4 Multi-user Activity Recognition

In this section, we first describe the basic reusable building blocks for multi-user HAR applications that are modeled according to the software abstractions and components of the proposed architecture. Then, we introduce a use-case on group hiking activities, discussing how a specific target scenario could be developed atop them.

4.1 Building Blocks of the Programming Model

As shown in Fig. 4, the basic operations of multi-user activity detection and recognition among CBSNs can be defined as follows:

1. *Neighbor Detection:* it detects neighbor CBSNs among co-located people in a specific range.

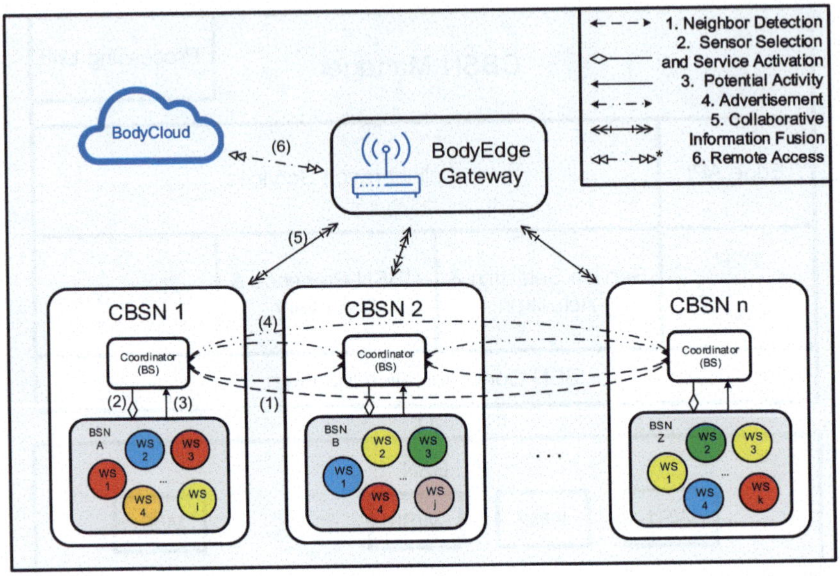

Fig. 4 The phases of multi-user detection process

2. *Sensor Selection and Service activation:* when neighbors have been detected, all of BSNs will activate/select sensing on their own nodes and necessary (processing) services will be activated. This step will allow the system saving energy which means sensor will only activated when it is needed.
3. *Potential Activity:* once a one-sided potential activity occurs on a BSN, the corresponding BS will be notified.
4. *Advertisement:* the coordinator will send a message to all neighbor CBSNs to request if there is a corresponding reaction for detecting multi-user activity.
5. *Collaborative Information Fusion:* low-level data and recognized individual activity will be sent to BodyEdge; the edge layer will perform decision-level fusion according to specific classification algorithms to detect the multi-user activity.
6. *Remote Access:* if large amounts of data are needed for computation or storage, the BodyCloud layer will provide proper support.

4.2 A Use-Case on Group-Hiking Recognition and Monitoring

In our use-case, we consider hiking activities performed by a group of people (e.g. tourists) with a professional guide. Therefore, we assume each excursionist will have a CBSN unit (see Fig. 5); the guide will carry, in addition, a mobile BodyEdge

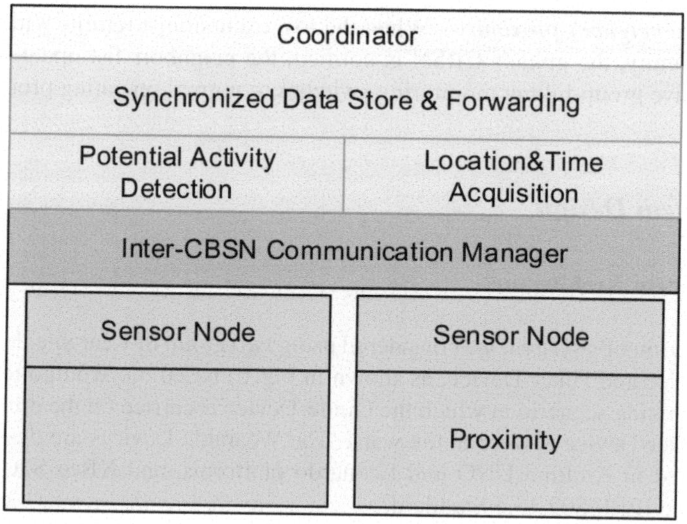

Fig. 5 The CBSN unit of the group-hiking use case

Gateway (e.g. a battery-powered Raspberry PI) inside the backpack. In this use case, it is not strictly necessary to be involve the Cloud layer so it will not be taken into account.

The CBSN employs the following two devices:

- The Smart Waist, which combines an embedded computing unit with motion sensor, is used to detect human's motion. From the motion sensor, several postures/activities (e.g., standing, walking, running) can be detected. These data will be processed via an Arduino and send to the coordinator.
- The Smart Coordinator (for Proximity acquisition), which provides location; it is used to track the out-door location.

The specific scenario taken in consideration is the separation of an excursionist from the rest of the group. When relevant physical activity transition is detected (e.g., a hiker stops to take a rest while the group keeps moving), there is possibility of separation from the group, which is typically a relevant event in group excursions.

Three possible situations can be detected by the periodic execution of the Neighbor Detection task:

- *Remaining within neighbor proximity*—the Potential Activity task will detect the transition (e.g. walk-to-stand), and notify to neighbor CBSNs by Advertisement task; then, it will activate the Proximity acquisition by Sensor Selection and Service activation function and geo-location to the BEG which starts the execution of Collaborative Information Fusion task to predict the separation event.
- *Leaving neighbor proximity*—the guide's CBSN detects that an excursionist has lost proximity, therefore the Smart Coordinator alerts the guide to support making a decision (stop and wait for the group member, slow down, or return to help).

- *Rejoining neighbor proximity*—When the lost excursionist returns within neighbor proximity, the guide's CBSN is notified, the neighbors list updated and the cooperative group-hiking monitoring switches to normal operating procedure.

4.3 System Design

4.3.1 System Architecture

In our experiment's system, we considered using two group of wearable devices (i.e. Guide-Device and Hiker-Device, as shown in Fig. 6) based on Arduino to simulate our group-hiking scenario in which the Guide-Device is carried on the guide's back, and the Hiker-Device is worn at the waist. The Wearable Devices are designed and implemented in Arduino UNO and Leonardo platforms, and XBee S2C modules based on the IEEE 802.15.4 Standard.

To simplify the scenario, we integrated the two different smart objects mentioned in Sect. 4.2 as Hiker-Device. The 3-axis accelerometer will achieve the same functionality as smart waist to recognize excursionist's activities such as walking and standing. The experiments in Indoor and Outdoor environments were considered scenarios free of external RF interference that operates in 2.4 GHz ISM band such as the XBee S2C modules or low interference power close to noise floor (around –100 dBm). Each XBee module was configured on the same 2480 MHz central operation frequency on channel 26 in IEEE 802.15.4 standard (16 channels are available, numbered from 11 to 26) to avoid overlapping with other wireless networks that operate on channel 13 like IEEE 802.11b/g standards. The XBee modules have a receiver sensitivity equal to –100 dBm (in normal mode on channel 26), i.e., an approximate noise floor of –100 dB.

As shown in Fig. 6, we have two different kind of devices:

- *Hiker-Device*—a 3-axis accelerometer (LIS344AL, acceleration range $\pm 3.5g$) was connected to the DFRduino Leonardo with Xbee Socket (Arduino Compatible) to capture user's motion data. An Xbee wireless module (featuring long range and low power consumption), will used to transmit data. Through the Xbee module, all the information will be sent to the Guide-Device.
- *Guide-Device*—an Xbee module is used to receive data transmitted by the End-Device and obtain the RSSI value when current RF packet received; then, all data was sent to the Arduino board, which provided its computation resources, to calculate and process data; the processed data was saved in a micro 2GB SD card.

4.3.2 Proximity Indicator

In a previous work [29], we reported that using uniquely inertial sensors is hard to disclose whether different subjects are in the same group while they have the same

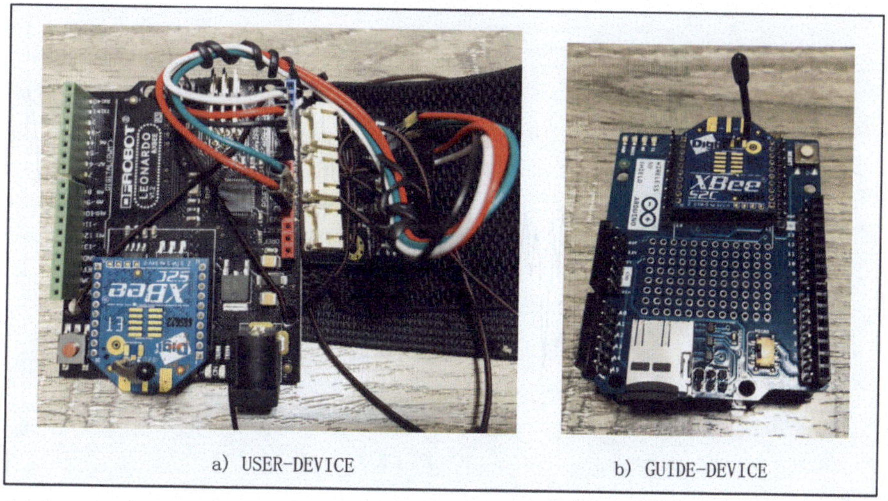

Fig. 6 Devices of the system (**a** is the hiker-device, and **b** is the guide-device)

acceleration pattern; therefore, in this paper we are more interested in using proximity indicators to tackle this issue.

Received signal strength indicator (RSSI), which is a measurement of the power present in a received radio signal, is used as the proximity indicator.

The distance d can be calculated from RSSI with the following formula, where A is the measured 1 m RSSI, and N is a parameter raging from 2 to 4 depending on the environmental factor.

$$d = 10^{\frac{A - RSSI}{10N}}$$

Usually, the RSSI signal has noise and other inaccuracies, therefore we smooth the signal with a Kalman filter. As shown in Fig. 7, the blue line is the raw data captured directly by the XBee module, and the red line is the filtered data.

4.4 Experiments and Results

We conducted experimental sessions with subjects that were asked to perform the activities reported in Table 1. We are interested in monitoring group activities such as group walking, leaving the group, and joining to a group. The Guide-Device was carried by one of the excursionists on the back, and the Hiker-Device was worn on the waist.

Each activity is repeated ten times and it is kept for 10 s. After written consent of each participant, the experiments were video recorded and manually labeled to have an accurate ground truth for comparing the output of algorithm.

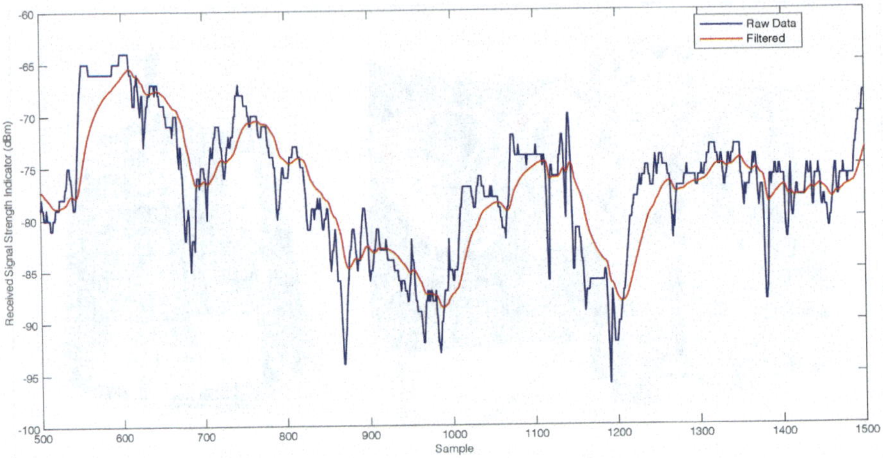

Fig. 7 Smoothed signal with Kalman filter

Table 1 Activity performed by the subjects

Activity	Description
Standing face to face	The subjects performed activity as standing face to face (distance = 2 m)
Standing back to face	The subjects performed standing activity and one of the subject stands face to the other's back (distance = 2 m)
Together	The subjects performed activity as walking together
Leave	One of the subjects moves away from the group
Join	One of the subjects gets close to the group

4.4.1 Raw Data Analysis

Figure 8 shows the activity with subjects face to face and back to face. When the subjects are back to face (i.e., guide in front and hiker following), there are no obstacles between the devices. When the subjects are face to face, their body is an obstacle in the wireless communication between the devices. In Fig. 8, we can observe that when there is human's body obstructing the signal, the average RSSI (red line) is smaller than without obstacle (blue line). The magnitude changes of the signal with line-of-sight transmission are also smaller than with obstacles between devices.

As shown Fig. 9, the hikers and the guide were together with their devices (all experiments have the same distance between subjects, i.e., 1 m). The RSSI changes around the average RSSI, and the fluctuation range is not large (around 5–15 dBm).

Also, we can observe that the signals in the figure are not stable; this is mainly due to two reasons:

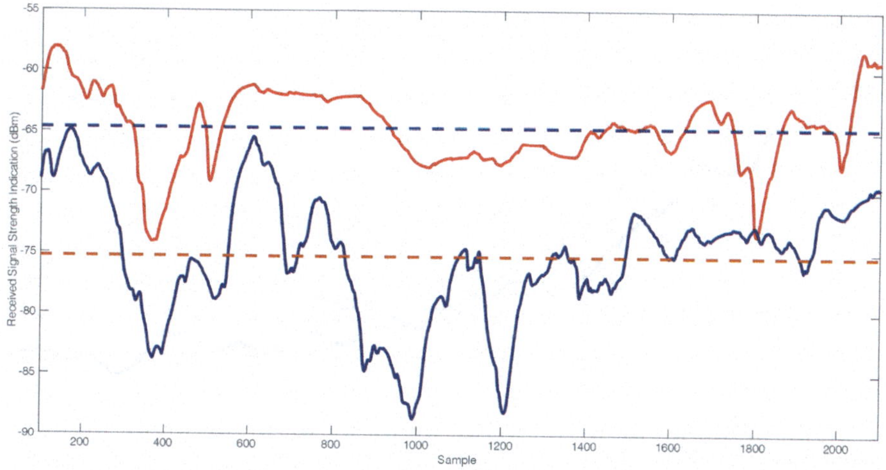

Fig. 8 Subjects standing face to face and back to face. Blue line is the average RSSI of signal with line-of-sight transmission; Red line is the average RSSI of signal obstructed by human body

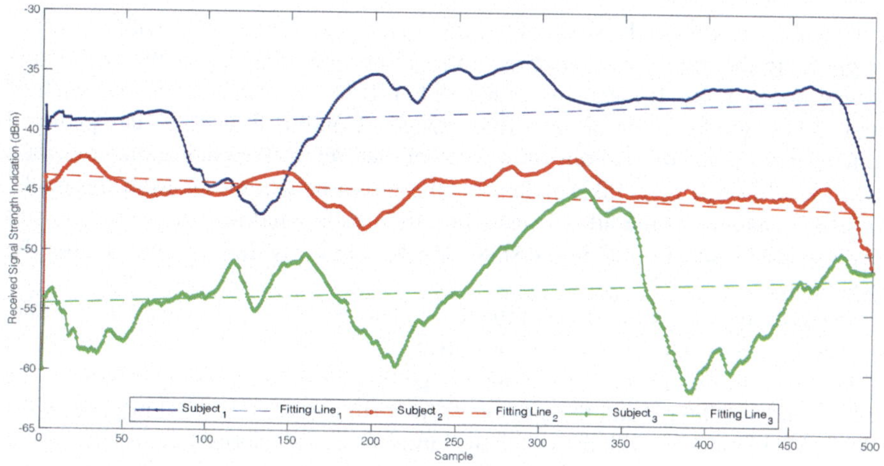

Fig. 9 RSSI curve of 'walking together' with fitted trend line

- when subjects are walking together, one of the subjects sometimes may go faster/slower than the guide; this phenomena will lead the devices communication shift from directly to indirectly (or from indirectly to directly), and let the signals become not stable and let the different subjects have different average RSSIs;
- the RSSI itself is not stable enough.

When subjects are walking together, one may sometimes go faster or slower than the others; this affects the communication link because of obstacles (i.e. human body) between the devices, making the signal unstable.

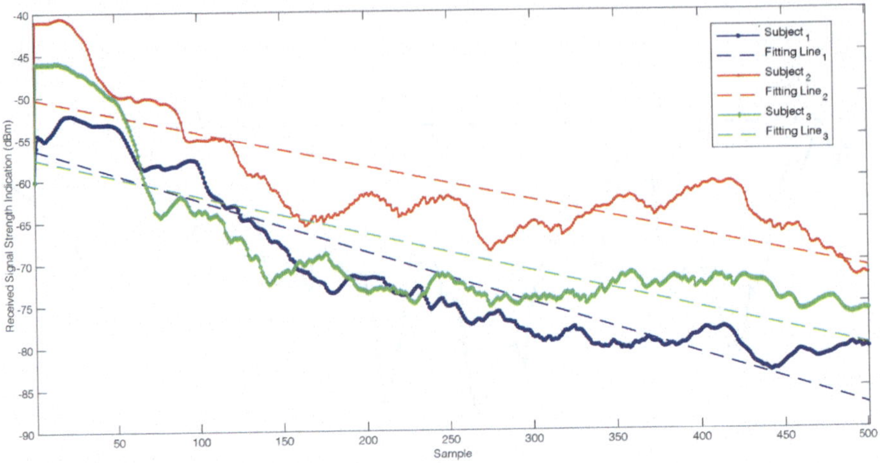

Fig. 10 RSSI curve of 'Leave' event with fitted trend line

The dash lines show the slope of the RSSI changes of three hiker with respect to the guide. In the figure, we can observe these lines are relatively stable and almost moving horizontally. It means in a time slide (10 s), the distance between each of them and the guide has no or very little changes. Combined with the data gathered from the motion sensors, we can know whether they are walking or standing together.

In Fig. 10, one of the hiker is leaving the group or getting far from the guide. The RSSI changes as the hiker leaves. In a time slide windows, the value tends to decrease with respect to the initial state. The further away they are, the smaller the RSSI value.

The dash lines show the slope of the RSSI changes that are obviously bigger than walking together, and the slopes are negative.

As shown in Fig. 11, one of the hiker is getting close to the group or guide. The RSSI changes as hikers and the guide get closer. In a time slide windows (10 s), the value tends to increase with respect to the initial state. The closer they are, the higher the RSSI value.

The dash lines show the slope of the RSSI changes that are obviously bigger than walking together, and the slopes are positive.

From the Figs. 9, 11, 10, we can observe that the trend of these three kinds of activities are different. Therefore, the slope of RSSI curves can be used to distinguish different activities. As shown in Fig. 12, the red Line shows the slope is usually below 0; the Blue Line is often above 0; the Green Line is around 0. From this information, we can easily define thresholds to distinguish different situations (i.e. leave group, join group, hike together).

With the Information above, we collected data under the three situations (i.e. Together, Leave, and Join) with multiple hiker/guide couples to empirically determine the thresholds. In a time slide window (10 s), as shown in Eq. (1), when $k \geqslant 0.02$, the distance between hiker and guide increases, and the hiker performs *Leave* activity;

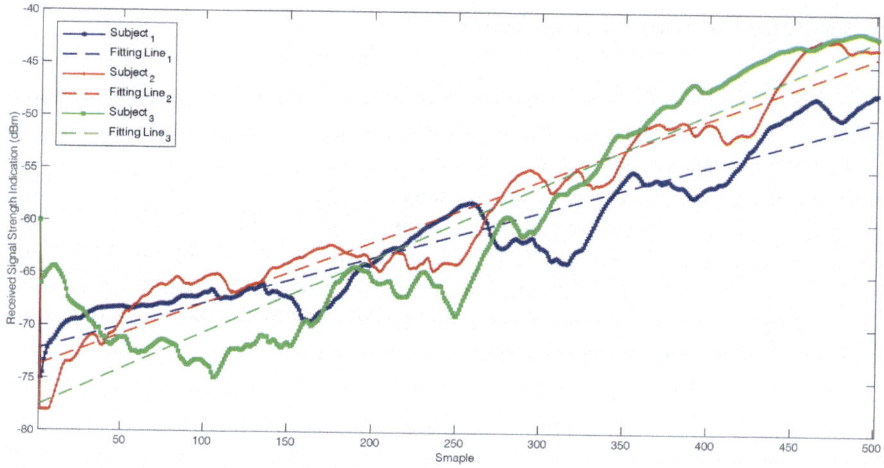

Fig. 11 RSSI curve of 'Join' event with fitted trend line

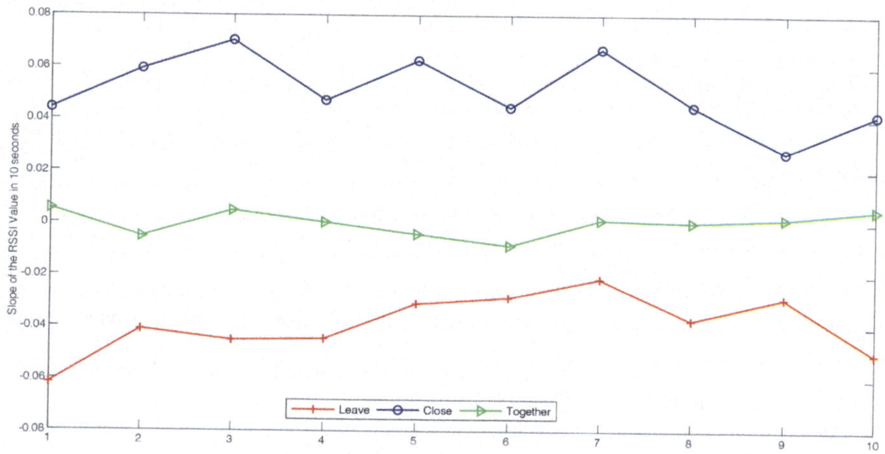

Fig. 12 Slope of the RSSI curves for Leave, Join, and Together situations

when $-0.02 < k < 0.02$, the distance between hiker and guide is stable, and the hiker performs *Together* activity; when $k \leqslant -0.02$, the distance between hiker and guide decreases, and the hiker performs *Join* activity.

$$Activity_i = \begin{cases} Leave & k \geqslant 0.02 \\ Together & -0.02 < k < 0.02 \\ Join & k \leqslant -0.02 \end{cases}, \ i = \{Leave, Together, Join\} \tag{1}$$

5 Conclusion and Future Work

In this paper, we proposed a preliminary definition of an architecture for supporting collaborative services based on BSN systems. The proposed architecture combines BodyCloud and BodyEdge with CBSNs to recognize group collaborations between multi co-located users. As use-case, we modeled specific events typical of hiking excursions. Using the proximity indicator, hiking together, hiker leaving the group, and hiker (re)joining the group can be easily classified. While ongoing efforts are devoted to complete the implementation of the use-case, we are planning to extend the architecture to support context-awareness, dynamic CBSN group self-definition and user-defined data privacy policies.

References

1. Lucero, S.: IoT platforms: enabling the Internet of Things. White paper. Accessed 15 Jan 2016
2. Rives, N.W., Serow, W.J., Rives Jr., N.: Introduction to Applied Demography: Data Sources and Estimation Techniques, vol. 39. Sage, Newbury Park (1984)
3. Lara, O.D., Labrador, M.A.: A survey on human activity recognition using wearable sensors. IEEE Commun. Surv. Tutor. (IEEE) **15**(3), 1192–1209 (2012)
4. Jalal, A., Kim, Y.H., Kim, Y.J., Kamal, S., Kim, D.: Robust human activity recognition from depth video using spatiotemporal multi-fused features. Pattern Recognit. (Elsevier) **61**, 295–308 (2017)
5. Wang, Z., Wang, J., Zhao, H., Qiu, S., Li, J., Gao, F., Shi, X.: Using wearable sensors to capture posture of the human lumbar spine in competitive swimming. IEEE Trans. Hum.-Mach. Syst. (IEEE) **49**(2), 194–205 (2019)
6. Ma, C., Gravina, R., Li, Q., Zhang, Y., Li, W., Fortino, G.: Activity recognition of wheelchair users based on sequence feature in time-series. In: 2017 IEEE International Conference on Systems, Man, and Cybernetics (SMC), pp. 3659–3664. IEEE (2017)
7. Gravina, R., Li, Q.: Emotion-relevant activity recognition based on smart cushion using multi-sensor fusion. Inf. Fusion (Elsevier) **48**, 1–10 (2019)
8. Ma, C., Li, W., Cao, J., Du, J., Li, Q., Gravina, R.: Adaptive sliding window based activity recognition for assisted livings. Inf. Fusion (Elsevier) **53**, 55–65 (2020)
9. Gravina, R., Ma, C., Pace, P., Aloi, G., Russo, W., Li, W., Fortino, G.: Cloud-based activity-aaService cyber–physical framework for human activity monitoring in mobility. Future Gener. Comput. Syst. **75**, 158–171 (2017)
10. Ma, C., Li, W., Gravina, R., Du, J., Li, Q., Fortino, G.: Smart cushion-based activity recognition and its typical application. IEEE Syst. Man Cybern. Mag. 2020 (Accepted)
11. Li, Q., Gravina, R., Li, Y., Alsamhi, S.H., Sun, F., Fortino, G.: Multi-user activity recognition: challenges and opportunities. Inf. Fusion (Elsevier) (2020)
12. Li, R., Chellappa, R., Zhou, S.K.: Learning multi-modal densities on discriminative temporal interaction manifold for group activity recognition. In: 2009 IEEE Conference on Computer Vision and Pattern Recognition, pp. 2450–2457. IEEE (2009)
13. Ibrahim, M.S., Muralidharan, S., Deng, Z., Vahdat, A., Mori, G.: A hierarchical deep temporal model for group activity recognition. In: Proceedings of the IEEE Conference on Computer Vision and Pattern Recognition, pp. 1971–1980. IEEE (2016)
14. Deng, Z., Vahdat, A., Hu, H., Mori, G.: Structure inference machines: recurrent neural networks for analyzing relations in group activity recognition. In: Proceedings of the IEEE Conference on Computer Vision and Pattern Recognition, pp. 4772–4781. IEEE (2016)

15. Kim, E., Helal, S., Cook, D.: Human activity recognition and pattern discovery. IEEE Pervasive Comput. (IEEE) **9**(1), 48–53 (2009)
16. Robertson, N., Reid, I.: A general method for human activity recognition in video. Comput. Vis. Image Underst. (Elsevier) **104**(2–3), 232–248 (2006)
17. Ribeiro, P.C., Santos-Victor, J., Lisboa, P.: Human activity recognition from video: modeling, feature selection and classification architecture. In: Proceedings of International Workshop on Human Activity Recognition and Modelling, pp. 61–78. Citeseer (2005)
18. Fortino, G., Parisi, D., Pirrone, V., Di Fatta, G.: BodyCloud: a SaaS approach for community body sensor networks. Future Gener. Comput. Syst. (Elsevier) **35**, 62–79 (2014)
19. Pace, P., Aloi, G., Gravina, R., Caliciuri, G., Fortino, G., Liotta, A.: An edge-based architecture to support efficient applications for healthcare industry 4.0. IEEE Trans. Ind. Inform. (IEEE) **15**(1), 481–489 (2018)
20. Augimeri, A., Fortino, G., Galzarano, S., Gravina, R.: Collaborative body sensor networks. In: 2011 IEEE International Conference on Systems, Man, and Cybernetics, pp. 3427–3432. IEEE (2011)
21. Wang, L., Gu, T., Tao, X., Chen, H., Lu, J.: Recognizing multi-user activities using wearable sensors in a smart home. Pervasive Mob. Comput. **7**(3), 287–298 (2011)
22. Alhamoud, A., Muradi, V., Böhnstedt, D., Steinmetz, R.: Activity recognition in multi-user environments using techniques of multi-label classification. In: Proceedings of the 6th International Conference on the Internet of Things, pp. 15–23 (2016)
23. Shimmer Website. www.shimmersensing.com. Accessed April 2019
24. Prossegger, M., Bouchachia, A.: Multi-resident activity recognition using incremental decision trees. In: International Conference on Adaptive and Intelligent Systems, pp. 182–191. Springer, Cham (2014)
25. Abkenar, A.B., Loke, S.W., Zaslavsky, A., Rahayu, W.: GroupSense: recognizing and understanding group physical activities using multi-device embedded sensing. ACM Trans. Embed. Comput. Syst. (TECS) **17**(6), 1–26 (2019)
26. Fortino, G., Guerrieri, A., Bellifemine, F.L., Giannantonio, R.: SPINE2: developing BSN applications on heterogeneous sensor nodes. In: 2009 IEEE International Symposium on Industrial Embedded Systems, pp. 128–131. IEEE (2009)
27. Fortino, G., Guerrieri, A., Bellifemine, F., Giannantonio, R.: Platform-independent development of collaborative wireless body sensor network applications: SPINE2. In: 2009 IEEE International Conference on Systems, Man and Cybernetics, pp. 3144–3150. IEEE (2009)
28. Iyengar, S., Tempia Bonda, F., Gravina, R., Guerrieri, A., Fortino, G., Sangiovanni-Vincentelli, A.: A framework for creating healthcare monitoring applications using wireless body sensor networks. In: 3rd International Conference on Body Area Networks (BodyNets'08), p. 8 (2008)
29. Li, Q., Gravina, R., Qiu, S., Wang, Z., Zang, W., Li, Y.: Group walking recognition based on smartphone sensors. In: EAI International Conference on Body Area Networks, pp. 91–102. Springer (2019)

Collaborative Solutions for Unmanned Aerial Vehicles

Francisco Fabra, Julio A. Sanguesa, Willian Zamora, Carlos T. Calafate, Juan-Carlos Cano, and Pietro Manzoni

Abstract In the coming years, regulation changes and market pushing are expected to relax existing restrictions to UAV flights. In this new scenario, we expect to encounter a greater number of UAVs over our cities, citizens, and critical infrastructures. Such changes impose new requirements in terms of safety, coordination, and operations management that must be properly addressed. In this chapter, we provide an overview of the main challenges that UAVs of the vertical takeoff and landing (VTOL) type are still facing, detailing their key application areas. Afterward, we discuss some solutions where wireless communications between UAVs enable achieving advanced collaborative solutions such as flight coordination and collision avoidance.

F. Fabra · C. T. Calafate (✉) · J.-C. Cano · P. Manzoni
Universitat Politècnica de València, Camino de Vera S/N, Valencia, Spain
e-mail: calafate@disca.upv.es

F. Fabra
e-mail: frafabco@cam.upv.es

J.-C. Cano
e-mail: jucano@disca.upv.es

P. Manzoni
e-mail: pmanzoni@disca.upv.es

J. A. Sanguesa
General Military Academy, Universitary Center of Defense, Zaragoza, Spain
e-mail: jsanguesa@unizar.es

W. Zamora
Universidad Laica Eloy Alfaro de Manabí, Manta, Ecuador
e-mail: willian.zamora@live.uleam.edu.ec

G. Fortino et al. (eds.), *Data Science and Internet of Things*, Internet of Things,
https://doi.org/10.1007/978-3-030-67197-6_7

1 Introduction

Unmanned Aerial Vehicles (UAVs), commonly known as drones, are semi-autonomous or fully autonomous unmanned aircrafts that are endowed with cameras, communication equipment and embedded sensors. Nowadays, they are being used for an increasing number of tasks and applications that significantly differ from their original military purposes.

Regarding recreational uses by enthusiasts, we can see how these flying devices have experienced a huge sale increase worldwide. They have also found new uses, including drone racing championships, which are becoming widespread. The building of custom UAVs is now frequent by both home users and students in different stages of the educational process, from high school to university degrees.

Regarding professional applications, they have gradually moved from more established areas like aerial photography and video, to other more novel areas including border surveillance [5], precision agriculture [1, 13], thermal inspections, package delivery, and air taxis [15], among other uses. In addition, UAVs are gradually replacing helicopters for aerial recordings since they represent a much cheaper alternative. Nowadays, they are also being used to assist in emergency situations such as search and rescue or disaster scenarios [2, 24], where they can be deployed on demand to act as supporting nodes for communications, as they offer a wider communications range and better line-of-sight (LOS) features than ground infrastructures.

Despite research in the UAV field has caused this area to experience significant advancements in recent years, battery lifetime remains one of the greatest challenges. Current battery technology based on Lithium Polymer (LiPo) is able to offer significant storage and discharge capacity given a relatively low volume and weight. However, this technology has several issues regarding maximum discharge thresholds, or long-term storage problems if the charge is too high, in addition to problems when operating at subzero temperatures. Thus, different alternatives to the LiPo technology are being sought that are able to extend the battery capacity, while also mitigating the well-known issues associated to this technology. Among the possible options being developed we have graphene-based batteries, solid-state batteries, and lithium-metal batteries, all of them promising to significantly improve the performance of current LiPo batteries.

The widespread use of drones also involves a number of critical issues to be considered, such as security, safety and privacy [17], particularly when flying over cities where the consequences of any flight problem are usually severe as the risk of injuring citizens is high. To address these issues, researchers worldwide are struggling to make UAV flights safer. U-space [16] is an european initiative that attempts to make the traffic management of UAVs more secure and safe. Specifically, it seeks to facilitate any kind of routine mission by providing an appropriate interface with manned aviation and air traffic control in all classes of airspace, thereby achieving the ambitious Single European Sky (SES) goal. The SESAR Joint Undertaking [23] was set up in order to manage this large scale effort, coordinating and concentrating all EU research and development activities focused on Air Traffic Management. This way, it will gradually become possible to perform complex drone missions that are currently restricted, hence achieving a sustainable and robust European ecosystem that is globally interoperable.

1.1 Social Impact

The huge success of UAVs has led the society as a whole to become more aware about the pros and cons of the presence and potential impact of these multifunctional flying devices. While the benefits they are able to provide are obvious, improper uses by unskilled or careless users has raised fears concerning privacy (as aerial recordings are hard to detect), and also alarm regarding accidents. In fact, different studies have recently highlighted the potential damage UAVs can cause to commercial planes [20], both on their wings or the nose cone of the plane. Such fears have been further aroused by events like those taking place at Gatwick airport in December 2018, where the presence of drones near this airport caused hundreds of flights to be cancelled, affecting more than 120,000 passengers. In fact, this event was not the first of this kind, and in the United Arab Emirates similar events have led local research groups to collaborate with authorities to design systems that are able to detect and take control over such rogue UAVs, forcing them to land in a safe location.

In the research community, several authors are focusing their efforts at developing UAV detection systems. For example, Dond and Zou [8] propose visual detection combining foreground detection and online feature classification in an attempt to reduce the problems related with dynamic backgrounds. Other authors, such as Jang et al. [14], propose the use of Acoustic Detection, using an Euclidean Distance Based algorithm; in particular, their proposal exploits the characteristic that the sound generated by most UAVs has harmonic complex tones.

Moving our focus to the transportation field, in Dubai the first autonomous UAVs capable of carrying one human passenger to predetermined places in the city are already operating [25], allowing citizens to envision an exciting future as we gradually bridge the gap towards some science fiction concepts like flying cars. In fact, replacing our cars by their flying counterparts, and making use of the vast aerial space, could solve the ever-growing traffic congestion problems that large metropolitan areas around the world are currently facing.

At a smaller scale, advanced indoor navigation systems are being developed [4] so that, combined with reconfigurable UAVs, image analysis, and deep learning algorithms, will allow these devices to adapt to their environment seamlessly. This way, UAVs can benefit from advanced features like passing through narrow spaces, recognizing people, animals and objects, and performing a wide range of tasks that will hopefully make everyone's life easier, while boosting safety and comfort levels by, e.g., assisting workers, helping the elder, or anyone requiring immediate physical intervention.

1.2 Legislation

Most people remain unaware of the actual flight restrictions of the different countries regarding maximum flight altitudes, flights in urban areas or at night, flights near airports, or in other restricted areas, or over groups of people. All these flight operations are forbidden for non-professionals in most countries that have regulated the use of these devices, and the simplicity of buying such products contrasts with the complexity at getting accurate information about their use. Fortunately, the situation is slowly changing, and amateur users have nowadays access to smartphone applications (e.g. IcarusRPA [22]) that simplify the access to such information. For professionals, initial regulations were very restrictive but, gradually, operations in scenarios that were initially forbidden (night flights, flight near airports, flights in urban areas, flights over crowds) are now possible in different countries as long as adequate measures are taken to guarantee the safety of citizens. In the particular case of urban environments, or in similar scenarios where UAVs are flying over groups of people, additional safety features are typically required, like the presence of parachutes, or other security measures that avoid a free fall. In fact, SKYCAT, PARAZERO, Fruity Chutes, and MARS Parachutes are just some of the different companies now offering such features as upgrades to commercial UAV models.

UAVs will be part of society in the coming years, and for this reason it is essential that legislation goes hand in hand with technological development to improve their inclusion.

1.3 Chapter Organization

The remainder of this chapter is organized as follows: Sect. 2 discusses a sense and avoid solution for UAVs. Afterward, Sect. 3 presents some solutions to handle collaborative UAV swarms. In particular, Sect. 3.1 discusses some swarm applications; then, Sect. 3.2 introduces MUSCOP, a solution to support mission-based swarms, and Sect. 3.3 introduces FollowMe, a solution to support manually guided swarms. Finally, in Sect. 4, we present the main conclusions for this chapter.

2 Sense and Crash-Avoidance Solutions

In the coming years, UAVs are expected to become ubiquitous, flying around in urban environments, both indoors and outdoors. Such a massive number of UAVs raises new challenges that deserve scrutiny, like the automatic detection and avoidance of potential collisions between UAVs.

UAV flight safety is being tackled from different areas. Yet, there is a particular area were research is lacking: the availability of sense and crash-avoidance mech-

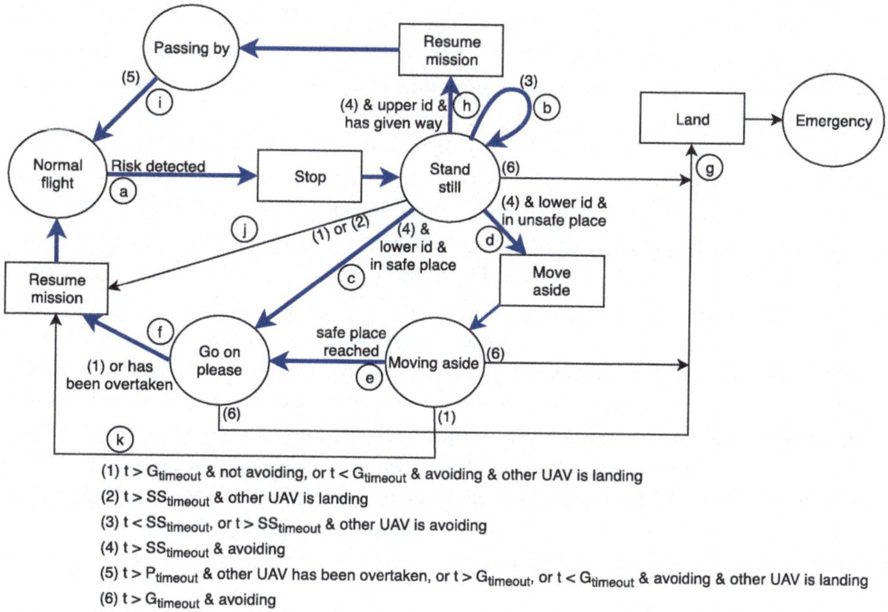

Fig. 1 Mission based collision avoidance protocol (MBCAP) finite state machine

anisms to allow UAVs to gain awareness of their environment, allowing them to elude other aerial vehicles if necessary [18]. To address this challenge, we proposed and implemented the Mission Based Collision Avoidance Protocol (MBCAP) [12], a solution that relies on wireless communications to detect other UAVs performing independent missions in a same area, and hence avoiding collisions with them. MBCAP works by making UAVs continuously broadcast their predictions regarding future positions, and stopping when they find out that their flight trajectories overlap in time and space with another UAV. In that case both UAVs involved will quickly negotiate to determine which UAV has a higher priority according to any specified criterium, and then execute the process to go safely through the critical area.

For a better understanding of how the MBCAP protocol works, Fig. 1 shows its finite state machine. The machine states are represented as circles, arrows represent the transitions between states, and rectangles represent the commands sent to the flight controller to change the behavior of the UAV. Blue arrows are indicative of transitions in the most common scenario where only two UAVs are involved.

We will proceed to describe in more detail this typical scenario where two UAVs meet, and a collision risk is detected. Since we assume that one of the UAVs will have more priority than the other, the situations handled by our protocol can fall into any of the following cases:

1. The *lower priority UAV* will start in the *Normal flight* state. Then, when detecting a collision risk, it stops in the air and enters the *Stand still* state (transition a). Afterward, a short time $SS_{timeout}$ (transition b) is required to guarantee that

the UAV with higher priority also reached that state. Upon being notified of the stop conditions by the other UAV, it checks whether the route of the high-priority UAV comes to near to its current position. If danger is detected, it changes its state through transition d, moving aside towards a new safe location; upon reaching it, transition e leads it to the $Go\ on\ please$ state. Otherwise, the UAV will directly switch to the $Go\ on\ please$ state (transition c), allowing the high-priority UAV to continue its mission as in the previous case. The low-priority UAV only resumes its mission (transition f) when the high-priority UAV has moved outside the conflicting area, which is an indicator that the collision has been successfully avoided.

2. The $higher\ priority\ UAV$ also starts its mission in the $Normal\ flight$ state, but switches to the $Stand\ still$ state (transition a) when detecting a collision risk. It will remain stopped (transition b) until the lower priority UAV allows it to resume its mission (transition h), upon which it moves to the $Passing\ by$ state. When the high-priority UAV detects that the distance towards the other UAV is starting to increase the overtaking process is considered to end, in which case it switches to the $Normal\ flight$ state (transition i), and notifies the low-priority UAV that it can resume its mission.

MBCAP was designed in a way to be resilient to unexpected events, and so additional transitions have been added. This means that, every time the UAV is in a state different from $Normal\ flight$, a global timeout is set, which elapses in two cases: (i) if the other UAV is close enough and the protocol has failed, in which case the UAV lands ($emergency$ state), and (ii) if the UAV fails to receive messages from the other contending UAV, in which case the mission is resumed (transitions f, i, j, k), as it considers that there is no longer a risk of collision. The only exception is the $Passing\ by$ state whereby the UAV resumes the mission instead of landing, as it considers that the low-priority UAV is not an obstacle, and so there is no actual risk of a collision.

In case a third UAV is involved, and it detects a collision risk with one of the UAVs that are solving a risky situation, it will stop (transition a), and wait in the $Stand\ still$ state (transition b) until the risky situation that was detected is solved. Later on, if necessary, the protocol is again triggered if the risk of collision with any of the UAVs persists.

We have validated the effectiveness of MBCAP through experiments with real UAVs, as well as through large-scale simulations. Our goal was to evidence the low overhead introduced both in terms of mission delays and wireless channel occupation.

Table 1 shows the performance achieved in an area sized 5×5 squared km when we vary the number of UAVs. We see that, indeed, our solution is quite efficient at avoiding collisions, even when the UAV density is high. A video summarizing our experiments is available online.[1]

Figure 2 provides a Google Earth 3D view that illustrates (i) with a red line, the path of the real multicopters, (ii) with a blue line, the path of the virtual high-priority

[1] https://youtu.be/bEdcsPX1hXY.

Table 1 MBCAP: success avoiding collisions (mean value by experiment)

Number of UAVs	25	50	75	100
Collisions expected	6.5	16.5	45.5	84.25
Risks detected	23.08	105.08	249.08	438
Collisions (*distance* < 4 m)	0.08	0.08	0.58	1.08

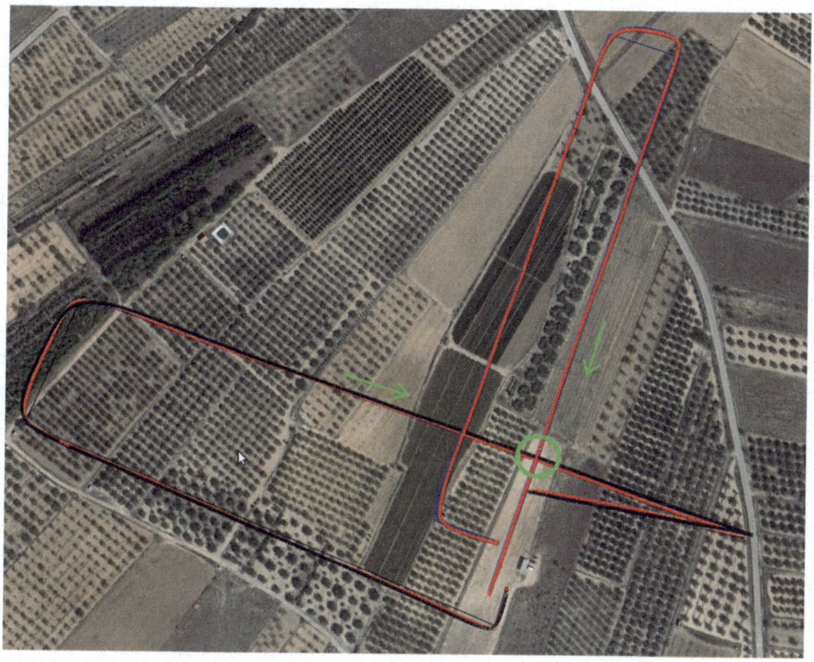

Fig. 2 Simulation versus real experiment in a perpendicular crossing

UAV and, (iii) with a black line, the route of the virtual low-priority UAV. The green arrows indicate the mobility direction for the UAVs upon detecting the collision risk at the location marked with a green circle. We see that both simulation and real paths present a high similarity. We have prepared a video showing these real experiments.[2] Please check [12] for more details on this protocol.

[2]https://youtu.be/xHnMuMOd9C0.

3 UAV Swarms

As highlighted in previous sections, UAVs pave the way for novel scenarios and applications. In these cases, UAV swarms have the potential to multiply their potential capabilities. In fact, autonomous UAV swarms are able to cooperate to monitor weather or traffic, transfer information, create or improve networks (e.g. solving infrastructure problems), or even accomplish distributed tasks [21], etc.

In general, the potential of the swarm increases as it becomes larger. Nevertheless, deploying large swarms requires handling very large amounts of data. Also, having more drones generates more input events that can have an impact on the behavior of the swarm, and could also increase the chances that drones crash into each other.

Various research teams worldwide are working on accomplishing complex tasks through an effective handling of UAV swarms. In 2016, Intel pioneered this area by creating the first UAV-based Light Show [19] using 500 drones. In December 2017, a similar show was created in China by the EHANG company, which used 1180 drones; in April 2018, that number increased to 1374 drones [7]. Then, in July 2018, Intel flew 2.018 drones to achieve the Guinness World Record for simultaneously flying the largest number of UAVs.

In general, the adoption of a swarm of UAVs can parallelize tasks by supporting the redundancy of different sensors, or by using different types of cameras simultaneously, and can also optimize some tasks through cooperation, among other scenarios.

3.1 Swarm Applications

Despite some solutions already exist for automating the flight of UAV swarms [6], automatic guidance may still be necessary in certain situations. Examples include applications for border surveillance [5], wild life recordings [3], or large-scale agriculture in search of pests or weeds [1], among others. When carrying out the mission, the UAVs that belong to the swarm must participated in swarm coordination tasks too. These missions must be planned in advance. Then, swarm consistency must be maintained through a near real-time responsiveness by relying on the communications between UAVs.

With respect to applications that rely on manual drone guidance, and were a pre-planned mission is not followed, swarm elements must adjust their routes dynamically to follow the master UAV, which is acting as the swarm leader. This solution may be required in scenarios such as disaster area monitoring, fire tracking, and search and rescue [2]. In such scenarios, the pilot must adapt the UAV heading in real time according to visual stimuli. In addition, some situations take place where, besides manual guidance, we need to carry multiple sensors or elements that exceed the lifting power of just one UAV. An example of such a scenario would include a rescue situation where multiple UAVs are used to carry medicine, water, food, or shelter.

3.2 Mission-Based Swarms

The reliability of communications is one of the main problems in the creation of swarms. A consistent distance must also be maintained by neighboring UAVs to avoid both collisions and communication disruption, which would hinder synchronization and cause the entire process to experience delays, or even reduce the actual number of UAVs conforming the swarm.

To deal with this problem we proposed the MUSCOP [11] protocol, which is a solution able to coordinate UAVs in order to maintain the target flight formation during planned missions. MUSCOP relies on a centralized approach where the master UAV synchronizes all the slave UAVs when an intermediate waypoint of a mission is reached. Our protocol allows different formations to be created around the leader (e.g. linear, matrix, circular).

The finite state machine which regulates the behavior of both master and slave UAVs is presented in Fig. 3. States are represented by circles, the messages sent and received are represented by curved arrows, and the transitions between states are represented by straight lines. On the left of the states we have curved arrows, which refer to the messages sent by the UAV ($TalkerThread$), while the messages received from other UAVs ($ListenerThread$) are represented through arrows on the right of the states.

Master and slave UAVs are identified by letters "M" and "S", respectively. Just before takeoff, the UAV that is positioned in the middle of the flight formation ("C") becomes the master, while the rest of the UAVs become slaves, being identified as "NC". It is worth pointing out that we are able to optimize communications by picking the UAV at the center of the formation as master, since most messages will be sent from master to slaves, and a few will be sent back from slaves to master; this way we are able to mitigate the impact of channel losses by making the sender/receiver distance become as small as possible.

Initially the UAVs are in state $Start$. When in that state, the master can determine the number of UAVs that will integrate the swarm flight formation as slaves must notify the master about their presence by sending it $hello$ messages. When all the UAVs have successfully connected to the master, it can switch to the $Setup$ state, which starts the initial configuration procedure. Initially, the location of the different UAVs in the flight formation are set by the master UAV. Then, according to the position of each UAV in the swarm, the master computes a set of waypoint coordinates, and sends a $data$ message with the results to each of the slaves, that details the specific mission each will have to follow. Then, the slaves send a $dataAck$ to the master informing that the message was received. Upon gathering confirmations from all the slaves, the master UAV switches to the $Ready\ to\ fly$ state, and forces the remaining UAVs to switch to that same state by broadcasting a $readyToFly$ message, and receives the corresponding acknowledgement ($readyToFlyAck$) in return. The master UAV starts the takeoff ($Taking\ off$ state) upon confirmation that all the slave UAVs are ready to fly as well, and thus will stop sending the $readyToFly$ message; such action forces slaves to take off as well. Only when all the UAVs reach their

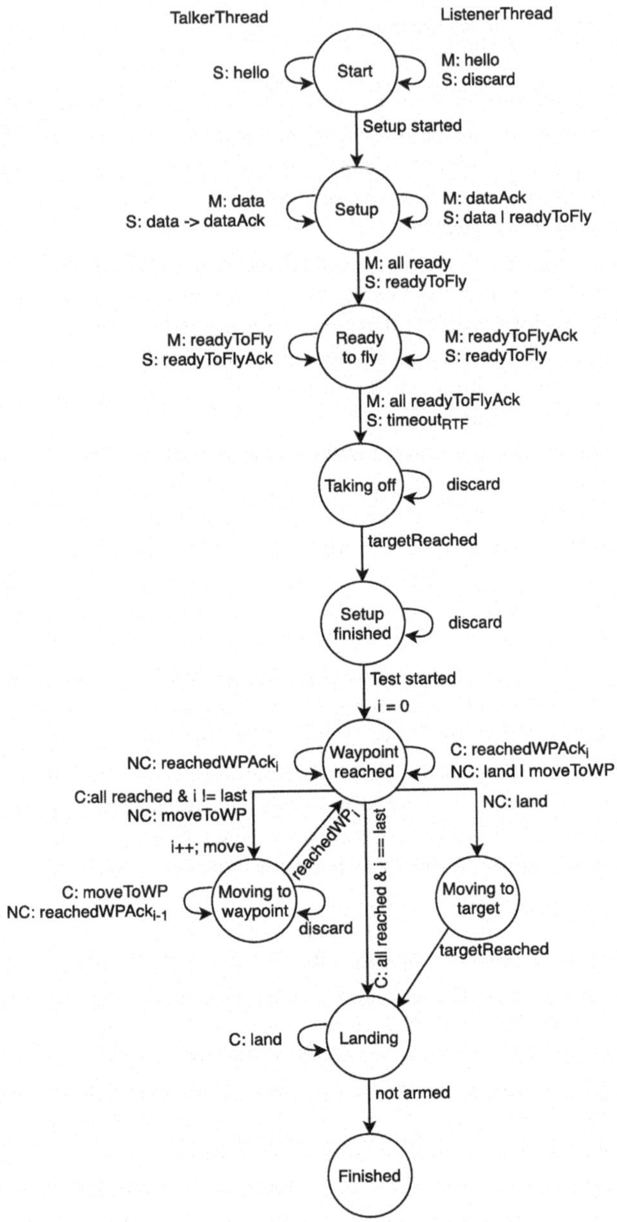

Fig. 3 MUSCOP protocol finite state machine

respective location in the flight formation does the setup process finish, switching to the *Setup finished* state.

When the swarm is ready, the user signals that the mission should begin. Since the UAV takeoff location is considered as the first waypoint of the mission, the UAVs will begin in the *Waypoint reached* state. When message *reachedWPAck* is received by the master UAV from all the slaves, the master starts moving towards the next waypoint (*Moving to waypoint* state), and it uses message *moveToWP* to also force slaves to move to the next waypoint. The different UAVs will remain in the *Moving to waypoint* state until the next waypoint is reached. While moving, the command to move to the next waypoint keeps being sent by the master UAV, and slave UAVs will merely acknowledge having reached the previous waypoint. Such redundancy helps to increase the level of reliability of our protocol, as distance or the presence of noise in the communications channel could make the messages sent by UAVs to be lost. Additionally, notice how our solution introduces a very low overhead on the wireless environment, since the messages sent are quite small (of respectively 6 and 14 bytes). When the master UAV detects that all the UAVs have reached the final waypoint of a mission, the master broadcasts the *land* message and proceeds to land at that location; the *land* message contains its location, and is used to force slaves to also initiate their landing procedure. However, we found that a mere landing of the swarm covers too much area, and would possibly make UAVs land on unexpected or hazardous locations, making it difficult to recover them. So, to optimize the landing process, the slaves first reduce the distance between them to just 5 m while keeping the same formation used during the flight, and only after do they land; this way we achieve a smaller landing area as intended.

We used the ArduSim [9] simulation framework to validate our solution. This simulator allows us to perform experiments in a realistic manner, and to validate the formations when varying the number of UAVs involved, and in two different wireless channels: ideal and lossy channel. We provide a video[3] that illustrates experiments with using three different flight formations (i.e. linear, circular and matrix) using ArduSim.

In terms of performance metrics, we measured the flight time when varying the number of waypoints (see Table 2), and we repeated the experiments five times, to obtain mean values. Then, we compared the flight time with the value obtained when a single UAV follows the same mission independently (*Reference mission*), obtaining the time overhead for the whole flight Δt. We found that the protocol only adds an overhead of about 0.55 s to the flight time per each waypoint crossed, which can be considered quite low.

To evidence the impact of channel quality on the swarm consistency, Fig. 4 illustrates the delay experienced (time offset) when we vary the distance separating the UAVs. Overall, we can see that a greater separation distance also causes the mean time offset to increase. We expected this since, in wireless channels, which are loss-prone, long distances will impair or harm the inter-UAV synchronization, which can become a critical issue. Also notice that, when the distance between sender and

---

[3] https://youtu.be/VLMsbL5B6tA.

Table 2 Flight time overhead

Number of waypoints	Reference mission (s)	MUSCOP mission (s)	Δt (s)
2	205.34	206.34	1.00
4	226.01	228.32	2.34
6	247.66	250.01	2.35
10	287.99	293.01	5.02
14	325.99	334.34	8.34
18	364.32	374.01	9.69
30	465.99	479.34	13.35

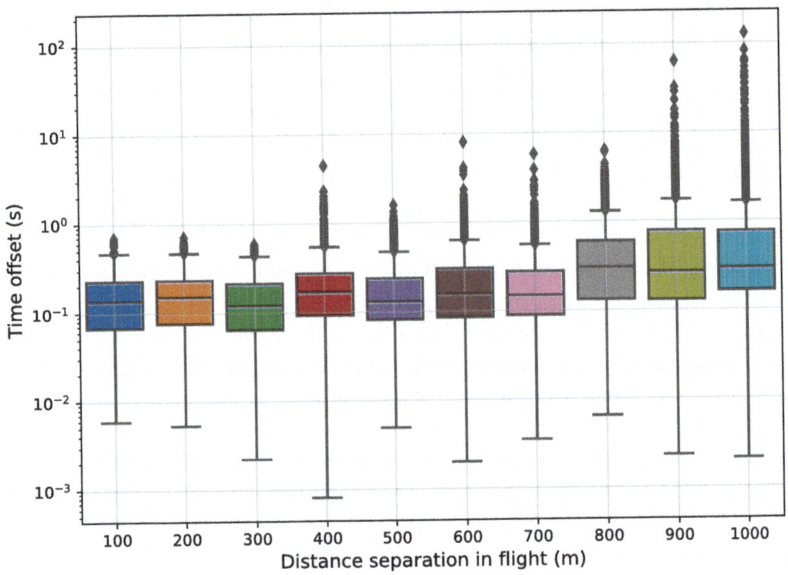

Fig. 4 Time offset when varying the inter-UAV distance

receiver UAVs is of less than 300 m, the time offset measured remains below 1 s, which we find to be quite acceptable.

3.3 *Manually-Guided Swarms*

Another solution able to set and keep a stable UAV formation is the FollowMe protocol we proposed in [10], and that is applicable to real cases where a professional pilot remotely controls the master UAV (swarm leader), while the remaining slave UAVs have to follow it in real time, and in an automatic manner.

The process works as follows: when the master UAV detects that all slaves are available and ready, it issues a message that includes its own coordinates, along with each slave's theoretical position to be adopted in the flight formation. When the master UAV is able to arrive at the same altitude as the slave UAVs in the formation, it will periodically broadcast a different type of message throughout the flight, that details its current location and its heading. Upon receiving such message, each slave calculates a new target location, and signals the flight controller so that the UAV proceeds to move to its expected location in the swarm.

The behavior of the master and slave UAVs is represented by the finite state machine shown in Fig. 5. The state sequence for the master UAV is presented above, and the state sequence for the slave UAVs is shown below. Curved lines represent the actions performed by the *Talker* thread, responsible for sending messages to other UAVs, and the lower lines represent the message that the UAV is waiting for at each specific state.

Initially the master UAV remains waiting for the flight controller to start accepting its commands ($UAVs\ ready$), which takes place in the $Start$ state. Afterward, slave UAVs in the $Wait\ mater$ state will signal their presence using message ID. Upon detecting all slaves, the $Setup\ started$ signal triggered by the user (using the setup button in the control app) makes it switch to the $Wait\ takeoff$ state. $TakeOff$ messages include all information required for slaves to determine their position in the target flight formation. When the master UAV receives message $Ready$ from all the slaves, confirming they have reached their formation location, it is ready to fly. At that stage, the pilot in command can start flying the UAV once the person controlling the flight presses the start button ($Test\ started$). When enough altitude is reached, it will switch to the $Follow\ me$ state, and at that time it will begin sending $Coordinates$ messages based on which slaves are able to follow the flight pattern of the leader. The flight will end when the master switches to the $Landing$ state, and sends slaves message $Land$ ordering them to end their flights. Notice that, in the entire process, the master waits for slave messages only during the first steps. As the master-slave model offers the master complete control over the remaining UAVs throughout the flight, the listener thread is of no use after the $Ready$ state is reached, and so it is dropped to save resources.

Slave UAVs also start the process by waiting in the $Start$ state until commands of the $UAVs\ ready$ type can be received by the flight controller. Afterward, these UAVs stay in state $Hello$, and will keep signaling their presence through ID messages until they receive a $TakeOff$ message from the UAV acting as master. The slaves will remain silent up to the moment when they arrive to their expected location in the formation, switching to the $Wait\ master$ state afterward, which means they are now prepared to follow the flight pattern adopted by the master ($Ready$ message). When a slave receives a command to follow the master UAV through the $Coordinates$ message, that slave will switch to the $Following$ state, and will keep following the master up to the time when they receive a $Land$ message. Since the talker thread is not needed once the UAV starts to follow the master ($Following$ state), that thread is terminated, similarly to what occurs for the listener thread following a land command.

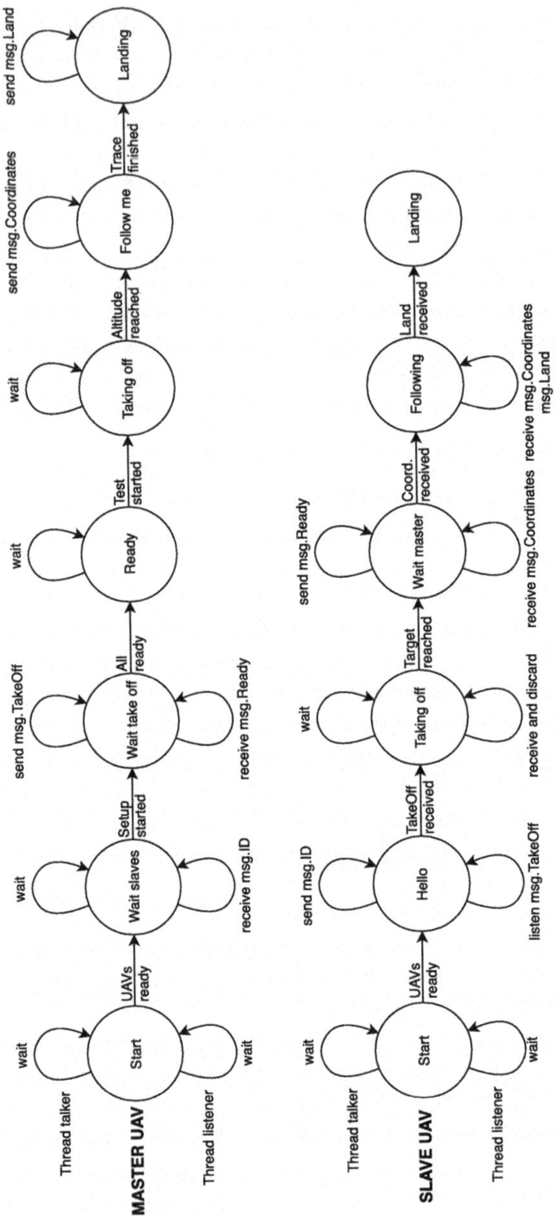

Fig. 5 FollowMe protocol finite state machine

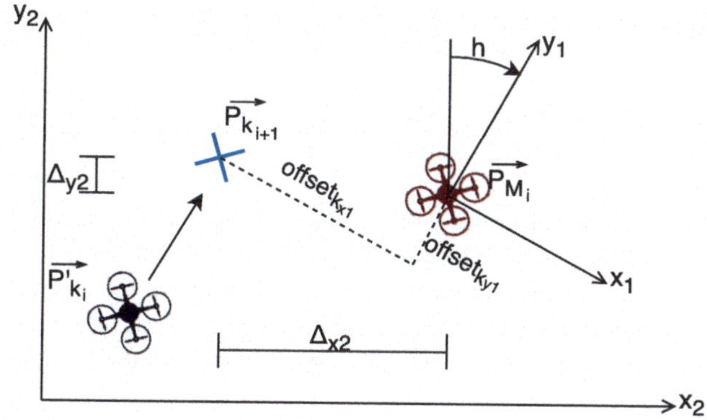

Fig. 6 FollowMe protocol operation

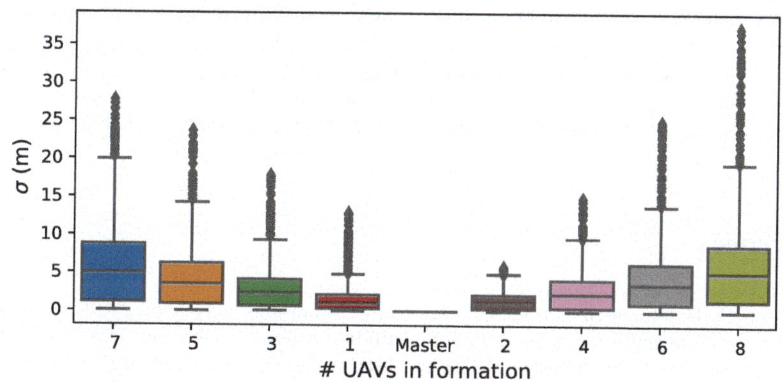

Fig. 7 Box and whisker plot of the formation error (σ) of the 9 UAVs of the swarm

During flight, the distance offset between the master and the target location of a specific slave is calculated considering the current heading of the master h (see Fig. 6), and therefore the slave constantly moves in an attempt to position itself at its specified location in the swarm formation $\overrightarrow{P_{k_{i+1}}}$.

Figure 7 shows how UAVs perform at maintaining swarm consistency depending on the separation from the leader considering a linear swarm layout. Notice that, when the distance towards the master becomes higher, the error values increase as well. Specifically, in a linear formation with 9 drones, slave drones number 7 and 8 show greater error than the UAVs that are closer to the master drone. This occurs because the master responds immediately to the commands issued by the pilot, and the remaining UAVs in the formation that have to follow the master will experience a delay that depends on the actual channel loss levels experienced.

4 Conclusions

As the complexity of the solutions targeted by UAVs continues to grow, so do the potential applications of these magnificent devices. In this chapter, we first introduced the main challenges and application areas for these flying devices. Afterward, we detailed some solutions where direct UAV-to-UAV communications are used to leverage collaborative solutions. In particular, we presented both sense and crash-avoidance and swarm management protocols that we have developed. These protocols have shown that it is possible to achieve acceptable or even excellent performance using existing technologies, although channel losses hinder performance to some extent.

Overall, we could say that UAVs represent nowadays one of the most exciting and promising research fields. In the upcoming years we will see if they are indeed able to meet the great expectations they have raised, and their actual social benefits.

Acknowledgements This work was partially supported by the "Ministerio de Ciencia, Innovación y Universidades, Programa Estatal de Investigación, Desarrollo e Innovación Orientada a los Retos de la Sociedad, Proyectos I+D+I 2018", Spain, under Grant RTI2018-096384-B-I00.

References

1. Albani, D., IJsselmuiden, J., Haken, R., Trianni, V.: Monitoring and mapping with robot swarms for agricultural applications. In: 2017 14th IEEE International Conference on Advanced Video and Signal Based Surveillance (AVSS), pp. 1–6 (2017). https://doi.org/10.1109/AVSS.2017.8078478
2. Aljehani, M., Inoue, M.: Multi-UAV tracking and scanning systems in M2M communication for disaster response. In: 2016 IEEE 5th Global Conference on Consumer Electronics, pp. 1–2 (2016). https://doi.org/10.1109/GCCE.2016.7800524
3. Anderson, K., Gaston, K.J.: Lightweight unmanned aerial vehicles will revolutionize spatial ecology. Front. Ecol. Environ. **11**(3), 138–146 (2013)
4. Balamurugan, G., Valarmathi, J., Naidu, V.P.S.: Survey on UAV navigation in GPS denied environments. In: 2016 International Conference on Signal Processing, Communication, Power and Embedded System (SCOPES), pp. 198–204 (2016)
5. Commission, E.: Autonomous swarm of heterogeneous RObots for BORDER surveillance. https://cordis.europa.eu/project/rcn/209949/factsheet/en. Accessed 30 Jan 2019
6. De Benedetti, M., D'Urso, F., Fortino, G., Messina, F., Pappalardo, G., Santoro, C.: A fault-tolerant self-organizing flocking approach for UAV aerial survey. J. Netw. Comput. Appl. **96**, 14–30 (2017). https://doi.org/10.1016/j.jnca.2017.08.004, http://www.sciencedirect.com/science/article/pii/S1084804517302606
7. Deng, I.: China's Ehang broke the world drone display record - but the aerial bots were out of sync (2018). https://www.scmp.com/tech/enterprises/article/2144393/chinese-start-broke-world-drone-display-record-aerial-bots-werent. Accessed 10 Jan 2020
8. Dong, Q., Zou, Q.: Visual UAV detection method with online feature classification. In: 2017 IEEE 2nd Information Technology, Networking, Electronic and Automation Control Conference (ITNEC), pp. 429–432 (2017). https://doi.org/10.1109/ITNEC.2017.8284767
9. Fabra, F., Calafate, C.T., Cano, J.C., Manzoni, P.: ArduSim: accurate and real-time multicopter simulation. Simul. Model. Pract. Theory **87**(1), 170–190 (2018). https://doi.org/10.1016/j.simpat.2018.06.009

10. Fabra, F., Zamora, W., Masanet, J., Calafate, C.T., Cano, J.C., Manzoni, P.: Automatic system supporting multicopter swarms with manual guidance. Comput. Electr. Eng. **74**, 413–428 (2019)
11. Fabra, F., Zamora, W., Reyes, P., Calafate, C.T., Cano, J., Manzoni, P., Hernandez-Orallo, E.: An UAV swarm coordination protocol supporting planned missions. In: 2019 28th International Conference on Computer Communication and Networks (ICCCN), pp. 1–9 (2019). https://doi.org/10.1109/ICCCN.2019.8847043
12. Fabra, F., Zamora, W., Sanguesa, J., Calafate, C.T., Cano, J.C., Manzoni, P.: A distributed approach for collision avoidance between multirotor UAVs following planned missions. Sensors **19**(10) (2019). https://doi.org/10.3390/s19102404
13. Faiçal, B.S., Pessin, G., Filho, G.P.R., Carvalho, A.C.P.L.F., Furquim, G., Ueyama, J.: Fine-tuning of UAV control rules for spraying pesticides on crop fields. In: 2014 IEEE 26th International Conference on Tools with Artificial Intelligence, pp. 527–533 (2014). https://doi.org/10.1109/ICTAI.2014.85
14. Jang, B., Seo, Y., On, B., Im, S.: Euclidean distance based algorithm for UAV acoustic detection. In: 2018 International Conference on Electronics, Information, and Communication (ICEIC), pp. 1–2 (2018). https://doi.org/10.23919/ELINFOCOM.2018.8330557
15. Johnson, W., Silva, C., Solís, E.: Concept vehicles for VTOL air taxi operations (2018)
16. M.I.C, E.A.S.A, C.A.A: Warsaw declaration: drones as a leverage for jobs and new business opportunities (2016). https://ec.europa.eu/transport/sites/transport/files/drones-warsaw-declaration.pdf. Accessed 3 April 2019
17. Mohamed, N., Al-Jaroodi, J., Jawhar, I., Idries, A., Mohammed, F.: Unmanned aerial vehicles applications in future smart cities. Technol. Forecast. Soc. Chang. (2018). https://doi.org/10.1016/j.techfore.2018.05.004, http://www.sciencedirect.com/science/article/pii/S0040162517314968
18. Mohammed, F., Idries, A., Mohamed, N., Al-Jaroodi, J., Jawhar, I.: UAVs for smart cities: opportunities and challenges. In: 2014 International Conference on Unmanned Aircraft Systems (ICUAS), pp. 267–273 (2014)
19. Nanduri, A.: Intel and drone technology – breaking new ground (2016). https://newsroom.intel.com/editorials/intel-and-drone-technology-breaking-new-ground/. Accessed 10 Jan 2020
20. Olivares, G., Lacy, T., Gomez, L., de los Monteros, J.E., Baldridge, R.J., Zinzuwadia, C., Aldag, T., Kota, K.R., Ricks, T., Jayakody, N.: UAS airborne collision severity evaluation. Technical report, National Institute for Aviation Research (2017)
21. Pace, P., Aloi, G., Caliciuri, G., Fortino, G.: A mission-oriented coordination framework for teams of mobile aerial and terrestrial smart objects. Mob. Netw. Appl. **21**, 708–725 (2016)
22. SKYDRONEX: IcarusRPA. https://www.icarusrpa.info. Accessed 10 Jan 2020
23. Undertaking, S.E.S.A.R.J.: SESAR JU. https://www.sesarju.eu/. Accessed 3 April 2019
24. Vincent, P., Rubin, I.: A framework and analysis for cooperative search using UAV swarms. In: Proceedings of the 2004 ACM Symposium on Applied Computing, SAC'04, pp. 79–86. ACM, New York (2004). https://doi.org/10.1145/967900.967919
25. Wakefield, J.: Dubai tests drone taxi service (2017). https://www.bbc.com/news/technology-41399406. Accessed 10 Jan 2020

Graph and Network Theory for the Analysis of Criminal Networks

Lucia Cavallaro, Ovidiu Bagdasar, Pasquale De Meo, Giacomo Fiumara, and Antonio Liotta

Abstract Social Network Analysis is the use of Network and Graph Theory to study social phenomena, which was found to be highly relevant in areas like Criminology. This chapter provides an overview of key methods and tools that may be used for the analysis of criminal networks, which are presented in a real-world case study. Starting from available juridical acts, we have extracted data on the interactions among suspects within two Sicilian Mafia clans, obtaining two weighted undirected graphs. Then, we have investigated the roles of these weights on the criminal networks properties, focusing on two key features: weight distribution and shortest path length. We also present an experiment that aims to construct an artificial network which mirrors criminal behaviours. To this end, we have conducted a comparative degree distribution analysis between the real criminal networks, using some of the most popular artificial network models: Watts-Strogats, Erdős-Rényi, and Barabási-Albert, with some topology variations. This chapter will be a valuable tool for researchers who wish to employ social network analysis within their own area of interest.

L. Cavallaro (✉) · O. Bagdasar
University of Derby, Derby, UK
e-mail: l.cavallaro@derby.ac.uk

O. Bagdasar
e-mail: O.Bagdasar@derby.ac.uk

P. De Meo · G. Fiumara
University of Messina, Messina, Italy
e-mail: pdemeo@unime.it

G. Fiumara
e-mail: gfiumara@unime.it

A. Liotta
Free University of Bozen-Bolzano, Bolzano, Italy
e-mail: Antonio.Liotta@unibz.it

G. Fortino et al. (eds.), *Data Science and Internet of Things*, Internet of Things,
https://doi.org/10.1007/978-3-030-67197-6_8

1 Introduction

Graph Theory is a well established field in mathematics. However, only recently many of its theoretical results started to be used within Social Network Analysis (SNA), an area with significant implications for real world scenarios. For example, one can simulate the behaviour of social networks using strategies like link predictions [1, 2], temporal networks, or spreading of influences [3, 4]. Other practical applications include to deal with large Artificial Neural Networks [5–7] or targeted advertisements to people based on their friends' interests [8] or, on the other side, containing the spread of fake news [9].

Network Science tools may also be used in the investigation of criminal networks. Sometimes the complex social interactions within a clan-based society may help the feature selection process for building machine learning models [10]. Other times, it is Network Science itself that helps conducting better performing investigation from law enforcement agencies. To this end, criminal networks can be encoded as graphs, and various types of analysis and simulations can be carried out for modelling criminal behaviours.

This chapter is to intended as a short tutorial on how Network Science strategies may be used to conduct an in-depth analysis on real criminal networks. Here, the Sicilian Mafia scenario has been considered. Section 4 relates to our previous analysis on this topic. In particular, Sect. 4.4 includes a comparative Degree Distribution analysis between real criminal networks and artificial ones. Indeed, when it is possible to find out a synthetic network reflecting the behaviour of real-world criminal networks, law enforcement agencies (LEAs) and network scientists can recreate those networks and simulate how interconnections among criminal will evolve.

This chapter is structured as follows. Section 2 presents the key theoretical tools (required for understanding the experiments conducted in Sect. 4), and it is divided in two parts: (i) tools, where the basic definitions on network science are provided; and (ii) popular artificial networks description (as the topologies used in Sect. 4.4). Next, in Sect. 3 a brief review on the use of (i) Social Network Analysis, and its implication in (ii) Criminal Networks is defined. Section 4 is a case study summarizing our work on two real criminal networks related to Sicilian Mafia [11, 12]. This section includes four parts: (i) datasets description, based on the data extracted from juridical acts; (ii) weights distribution analysis, which represents an important preliminary study to understand how the interactions among suspected are structured in terms of interaction frequency; (iii) shortest path analysis, which allows to identify trusted affiliates inside the clan who can spread confidential and illegal messages; (iv) comparative degree distribution analysis between real and synthetic networks, to artificially recreate the criminal networks used here, with the purpose of conducting further investigations through them. Finally, the conclusions follow in Sect. 5.

2 Complex Networks

In this section we introduce the main Network Science concepts underpinning Social Network Analysis, which are later exemplified in the criminal network case study (Sect. 4). In particular, in Sect. 2.1 the main definitions required for understanding the mechanics of SNA are provided. All theoretical concepts are derived from [13], which we refer to the reader for further technical details.

2.1 Tools

In this section, we start with some basic definitions of Graph Theory.

Graph

Definition 1 A *graph* denoted by $G = (N, E)$, consists of a set of nodes N and a set of edges $E \subseteq N \times N$ (also called links L). It is a convenient way of representing relationships between pairs of objects.
As an example, *Facebook®* may be viewed as a graph, where the nodes represent users and edges represent the friendship relationship among them. Is it also possible to define a *subgraph* as follows:

Definition 2 A *subgraph* H of the graph G is a graph whose nodes and edges are subsets of the nodes and edges of G.
Furthermore, graphs may be either *weighted*, or *unweighted*:

Definition 3 A *weighted graph* $G = (N, E, W)$ is a triplet consisting of a finite set of nodes N, a set of edges E, and a set of weights $W : E \rightarrow \mathbb{R}$ defined on each edge. If all edges weights are equal to one, then the graph is called *unweighted*.

Degree

Definition 4 The *degree* of a node n_i, denoted $deg(i)$ or k_i, is the number of incident edges to n_i. The sum of the degrees of all nodes is equal to the double of the number of edges E:

$$\sum_{n \in N} k_n = 2E. \tag{1}$$

Definition 5 In weighted networks, the *weighted degree* (also known as *strength* [14, 15]) is the sum of the edges weights w incident on n_i:

$$k_i = \sum_{(i,j) \in E} w_{ij}, \tag{2}$$

where the summation spans over all edges (i, j) in the network, linked to node n_i.

Definition 6 For undirected networks, the *average degree* is defined as

$$\langle k \rangle = \frac{1}{N} \sum_{i=1}^{N} k_i = \frac{2L}{N}, \tag{3}$$

where N is the total number of nodes, k_i is the degree of a generic node i, and L represents the total number of links, or edges E, within the network.

Small-World

The Small World phenomenon [16, 17] is based on the concept of the six degrees of separations, according to which two random people in the world may be connected each other via a few acquaintances (i.e., it is estimated that there are six people in the middle between the source and the destination). In Network Science, it translates into a "short" distance between two randomly chosen nodes within a network, that is

$$\langle d \rangle \approx \frac{\ln N}{\ln \langle k \rangle}, \tag{4}$$

where N is the total number of nodes in the graph, $\langle k \rangle$ is the network average degree, and $\langle d \rangle$ the average distance within the network. The denominator implies that the denser the network, the smaller the distance between the nodes is. In conclusion, the average path length or the diameter depends logarithmically on the system size.

Degree Distribution

The *degree distribution* p_k provides the probability that a randomly selected node in the network has degree k. Since p_k is a probability, it must be normalized; i.e.,

$$\sum_{i=1}^{\infty} p_k = 1. \tag{5}$$

For a network made of N nodes, the degree distribution is the normalized histogram given by:

$$p_k = \frac{N_k}{N}, \tag{6}$$

where N_k is the number of nodes having degree k.

The degree distribution has assumed a central role in network theory following the discovery of scale-free networks (See "Scale-Free Property" paragraph); moreover, p_k determines many network phenomena, from network robustness to the spread of viruses.

Weight Distribution

The degree distribution can be extended to weighted networks considering the weighted degree (strength) distribution $P(s)$, defined as the probability that a node may have weighted degree (strength) equal to s. Based on [15], this is

$$P(s) \sim s^{-\gamma}, \tag{7}$$

where γ is a constant typical of the network.

Clustering Coefficient

Clustering is used to quantify the relationship among nodes' neighbours. Indeed, the degree only considers the number of direct links between nodes. The *clustering coefficient* C_i measures the edge density in the immediate neighbourhood of a node. $C_i \in [0, 1]$ represents the clustering coefficient of a generic node n_i:

$$\begin{cases} \text{if } C_i = 0, & \text{there are no edges among the node's neighbours} \\ \text{if } C_i = 1, & \text{each node's neighbour is connected with the others} \end{cases}$$

The local clustering coefficient is computed as follow:

$$C_i = \frac{2L_i}{k_i(k_i - 1)}, \tag{8}$$

where k_i is the degree of the generic node n_i, and L_i represents the number of links (i.e., edges) between the k_i neighbours of n_i.

Average Clustering Coefficient

The average $\langle C \rangle$ of $C_i \in i = 1, \ldots, N$ in the whole network is given by

$$\langle C \rangle = \frac{1}{N} \sum_{i=1}^{N} C_i. \tag{9}$$

Adjacency Matrix

A common way to represent relationships among nodes is the *adjacency matrix* A.

Definition 7 The *Adjacency Matrix* $A[i, j]$ holds node degree (weighted or unweighted) to the edge (n_i, n_j) if it exists, where n_i is the node with index i and n_j is the node with index j. If there is no such edge, then $A[i, j] = None$.

For undirected graphs A is symmetric (i.e., $A[i, j] = A[j, i]\ \forall n_i, n_j \in N$).

Path

Definition 8 A *path* is a sequence of alternating nodes and edges that flow from a starting node to an ending one such that each edge is incident to its predecessor and successor node. A path is called *simple* if each node in the path is distinct.

More formally, a path can be defined as a sequence of nodes

$$P = (n_1, n_2, \ldots, n_m) \in N \times N \times \cdots \times N,$$

such that n_i is adjacent to n_{i+1} for $1 \leq i \leq m - 1$. Such a path P is called a path of length $m - 1$ from n_1 to n_m.

Measures based on paths strategies are the **shortest path length analysis**.

Distance

Definition 9 The *distance* from a node n_i to a node n_j in G, denoted $d(n_i, n_j)$ is the length of a **shortest path** from n_i to n_j (if such a path exists).

$$d_{ij} = min\,(\Gamma(i, j)),$$

where $\Gamma(i, j)$ is the set of paths connecting i and j.

Connectedness

Definition 10 A graph G is *connected* if, for any two nodes, there is a path between them.

Definition 11 If G is not connected, its maximal connected subgraphs are called the *connected components* of G.

Definition 12 If a network consists of two components, a properly placed link can connect them, making the network connected. Such a link is called *bridge*.

Scale-Free Property

The majority of real networks, such as the World Wide Web, are called *scale-free networks* and follow the definition:

Definition 13 A scale-free network is a network whose degree distribution follows a power law.

The *power-law distribution* has the following form

$$p_k \sim k^{-\gamma}, \tag{10}$$

where the exponent γ is its *degree exponent.*

Some artificial network models such as the **Barabási-Albert (BA) Model** successfully exhibit this feature.

2.2 Artificial Networks

The need for scientists to create Artificial, or Synthetic Networks has been born from the aim to reproduce real network properties in a controlled environment. For this reason, several typologies of Artificial Networks have been formulated.

Three models in particular have found special popularity within the scientific community: The Erdős-Rényi (ER, also known as Random Network) Model, the

Watts-Strogats Model (WS; i.e., a Random Network variation), and the Barabási-Albert (BA) Model. This last one tries to capture two important properties of real network: the growth and the preferential attachment. Further details on those models are provided in the following paragraphs.

Random Network Model

A random network consists of N nodes where each node pair (n_i, n_j), $\forall i, j \in N$ is connected with probability p. To construct a random network one needs to

1.Start with N isolated nodes,
2.Select a node pair (n_i, n_j) and generate a random number $rand \in [0, 1]$:

$$\begin{cases} \text{if}\, rand > p, & \text{connect the selected node pair with a link} \\ \text{otherwise,} & \text{leave them disconnected} \end{cases}$$

3.Repeat the previous step for all pairs of distinct nodes $(n_i, n_j) \in N \times N$.

The network obtained after this procedure is called a *random graph* or a *random network*. There are two definitions of a random network: the definition provided in the Erdős-Rényi Model, and the one of the Gilbert Model.

Erdős-Rényi Model
Random networks are also called *Erdős-Rényi Networks* from the names of the mathematicians Paul Erdős (1913–1996) and Alfréd Rényi (1921–1970), who studied the properties of these networks. Their model follow the structure

$$G(N, L), \tag{11}$$

where N labeled nodes are connected with L randomly placed links (i.e., edges). Paul Erdős and Alfréd Rényi used this definition in their paper [18].

Gilbert Model
It is a variation of the Erdős-Rényi Model. It has been defined by Edgar Nelson Gilbert (1923–2013) and follows the structure

$$G(N, p), \tag{12}$$

where each pair of N labeled nodes is connected with probability p.

There are two main limits in Random Network Model that had to be overcome over the years by the academic community:

1. The local clustering coefficient in ER model is given by [13]

$$C_i = \frac{\langle k \rangle}{N}.$$

   This behaviour of C_i is contradicted by the local clustering coefficient of real networks.
2. The Poisson distribution that describes the degree distribution of ER networks does not allow large differences between the worst- and best-connected nodes in the network. This implies that hubs, frequently observed in real networks, cannot be found in ER networks. BA model, relying on preferential attachment and growth, successfully reproduces this fundamental feature.

In the following paragraphs those models are described.

Watts-Strogats Model

Two main considerations motivated Duncan J. Watts (1971) and Steven Strogatz (1959) to propose this model: (i) in real networks the average distance between two nodes depends logarithmically on N (See "Small-World" average distance $\langle d \rangle$); (ii) the average clustering coefficient $\langle C \rangle$ of real networks is much higher than expected for a random network of similar N and L (i.e., E).

To construct a random network according to *Watts-Strogats* Model [19]:

1. Start from a ring of N nodes, whereas each node is connected to its immediate previous and next neighbours; hence, each node has $\langle C \rangle = 3/4$, initially.
2. With probability $p \in [0, 1]$ each link is rewired to a randomly chosen node

$$\begin{cases} \text{if } p \simeq 0, & \text{regular lattice} \\ \text{if } 0 < p < 1, & \text{Small-World property} \\ \text{if } p = 1, & \text{Random Network Model (all links rewired).} \end{cases}$$

The Watts-Strogatz model interpolates between a *regular lattice*, which has high clustering (but lacks the Small-World phenomenon), and a *random network*, which has low clustering (but displays the Small-World property). Moreover, high nodes degrees are absent from Watts-Strogatz model.

Barabási-Albert Model

This model was theorized by Albert-László Barabási (1967) and Réka Albert (1972) [20]. It simulates a Scale-Free Network rather than a Random one, by introducing two new concepts to the model: network growth and the preferential attachment. The first concept assumes that real networks continuously increase over time, so new nodes must be considered. The second point argues that random connections defined by a fixed probability do not reflect the behaviour of real networks; in fact, in real scenarios, nodes tend to link to the more connected nodes.

Definition 14 The *Preferential Attachment* is the probability $\Pi(k)$ that a link of the new node n_j connects to node n_i depends on the degree k_i through the formula

$$\Pi(k_i) = \frac{k_i}{\sum_j k_j}. \quad (13)$$

To construct an artificial network with the Barabási-Albert model, the steps are:

1. Start with a set of N_0 nodes, the links between which are chosen arbitrarily, as long as each node has at least one link.
2. **Growth** – At each timestep a new node n_j with l links (with $l \leq l_0$) that connects the new node to nodes already in the network is added.
3. The connections between the new node with the older nodes are defined by the **Preferential Attachment** probability.

3 Social Network Analysis in Criminal Networks

In this section we provide an overview of state-of-the-art of Social Network Analysis applied to Criminal Networks. We also consider the most relevant studies concerning specifically the Sicilian Mafia criminal topologies.

3.1 Criminal Networks Analysis

Through SNA, LEAs are able to analyze criminal networks and investigate the relations among criminals. For this reason, nowadays there is a growing interest in the application of Graph and Network Science onto criminal networks. For instance, SNA has been used in [21] to build crime prevention systems. However, due to the lack of data availability on those kind of networks, there are difficulties in finding relevant quantitative studies. Such examples are those conducted by Szymanski [22] and Berlusconi [23], on the problem of community detection and link prediction.

3.2 Sicilian Mafia Networks

Sicilian Mafia has a particular structure that differs from common criminal networks (such as the terrorist nets), whereby it is a common practice for criminals to come together to achieve a common goal and then fall apart. By contrast, in Sicilian Mafia this behaviour does not occur. Indeed, the affiliates are bound by blind loyalty and they still pursue further goals even after achieving a previous one. Moreover, Families last for several generations. They also tend to diversify their objectives: from controlling entire economic sectors (e.g., by giving "protection" to small traders and taking control of larger factories), to influencing countries political life (e.g., by interfering in the results of electoral competitions). Sicilian Mafia originated in Sicily, and has now spread worldwide [24–26]. The blind loyalty of affiliates makes it even more difficult to obtain reliable information about those criminal networks topologies: important information about such criminal network is likely to be missing or hidden, due to the covert and stealthy nature of criminal actions [27–30].

4 Case Study: The Sicilian Mafia

This section describes firstly the real criminal datasets we used for our tests, followed by a brief summary on the strategies conducted jointly with the results obtained so far as an example on Network Science strength, and how it can be used to significantly help LEAs. In particular, the experiments relate to: (i) weight distribution, (ii) shortest path length, and (iii) degree distribution.

4.1 Dataset Description

The case study example relates to two real-world datasets [31] we built from juridical acts[1] [12]: (i) the *Meetings* dataset represents the physical meetings among criminals obtained through LEA evensdropping; (ii) the *Phone Calls* dataset refers to phone calls between individuals obtained through LEA interceptions.

This particular investigation was a prominent operation conducted during the first decade of the 2000s and focused on two Mafia clans known as the "Mistretta" family, and the "Batanesi" clan [11, 12, 32].

Both datasets led to undirected and weighted graphs; thus, edge weights w are available and represent the number of times any given pair had a meeting in the *Meetings* dataset, and the number of times two individuals called each other in *Phone Calls* dataset. In SNA, those coefficients are also known as the strength of the tie binding two individuals [11]. In Fig. 1, the graphs obtained as well as the description of what each element represents (i.e., nodes, colours, edges weight, nodes and edges size, etc.) is shown. The main characteristics of the datasets are summarized in Table 1.

[1] Source code are available at https://github.com/lcucav/criminal-nets.

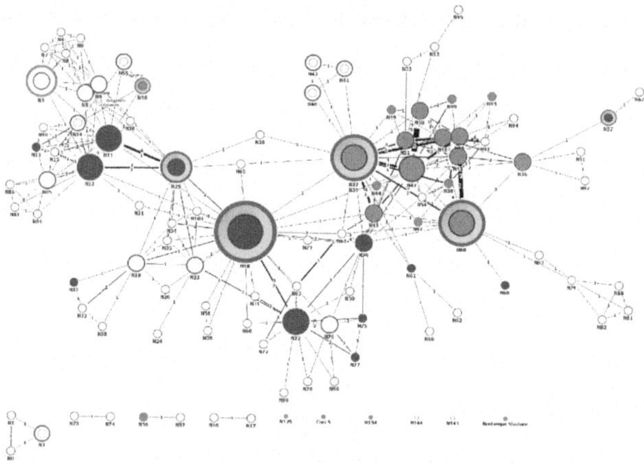

(a) MEETINGS

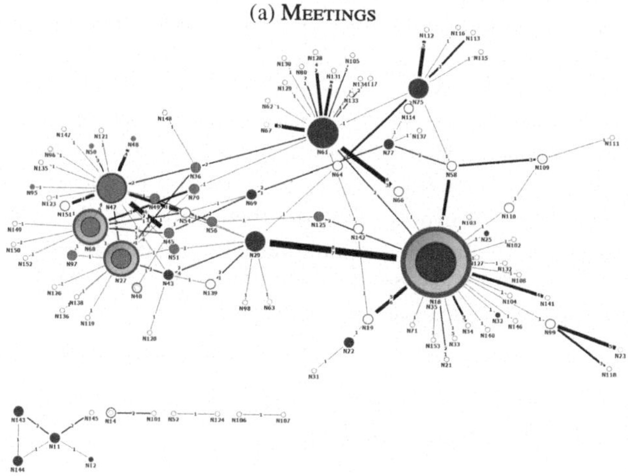

(b) PHONE CALLS

Fig. 1 Dataset Description. The colours represent different clans: darker nodes are the "Mistretta" family; in grey the "Batanesi" clan is drawn; white and light gray circled nodes and for two others Mafia families not directly involved in the current investigation. All *circled* nodes represent the *bosses*. Lastly, white nodes represent other subjects not classifiable in any of the previous categories. Edges' width depends on the number of meetings or phone calls, while the nodes size relates with their degree. (Reproduced from Ficara et al. 2020)

4.2 *Weight Distribution Analysis*

Figure 2 shows the weight distribution of the *Meetings* and the *Phone Calls* networks. As already mentioned, the weights represent the amount of meetings and phone calls exchanged between pairs of individuals in the networks, respectively.

It is noteworthy that in both these networks there are just a few high-weight edges; i.e., nodes incident on those links exhibit an high number of interaction within the

Table 1 Characteristics of *Meetings* and *Phone Calls* networks. (Reproduced from Ficara et al. 2020)

Parameter	Meetings	Phone calls
No. nodes	101	100
No. edges	256	124
Max. weight	10	8
Max. frequency	200	100
Avg. degree	5.07	2.48
Max. shortest path	7	14
Common nodes	47	

network. In [11], we motivated this behaviour as a necessity from affiliates to focus their efforts in trying to reduce the risk of being intercepted by external people (i.e., LEA, and other people outside the clan). In the *Meetings* network, this trend is even more accentuated; moreover, the maximum interactions weight (i.e., $w = 10$) is greater than its counterpart in the *Phone Calls* network (i.e., $w = 8$).

Our explanation is that mobsters prefer to communicate by face-to-face meetings, rather than calling each other, to reduce interception risks. Furthermore, bosses often have to participate to public events to pursue their power inside a clan, including: funerals of other affiliates, and other solemn religious demonstrations (masses, processions, etc.). It is a well known practice that, during those kinds of events, bosses pass messages to their closest subordinate affiliates.

4.3 Shortest Path Length Analysis

The shortest path length distribution in Fig. 3 is closely related to dynamic properties such as velocity of messages spreading process within the network. Generally speaking, the criminal organizations structure aims to optimize the interaction frequency among members, while reducing as much as possible the interception risks. Thus, trusted members may be discovered by following short interactions paths; indeed, those affiliates may also be acting as a *bridge* (See Sect. 2.1"Connectedness") to connect distant groups in the network.

In [11] we noticed that both the weighted and the unweighted shortest path length analyses show a higher interaction frequency among affiliates throughout a "balanced" number of intermediates. This means that they do not like to spread their encrypted messages with a too low (resp., high) number of intermediates. This is to avoid, from one side, to overexpose their bosses to police investigations. From the other side, the longer the sequence of intermediates, the higher the chances to be intercepted by people outside the Family.

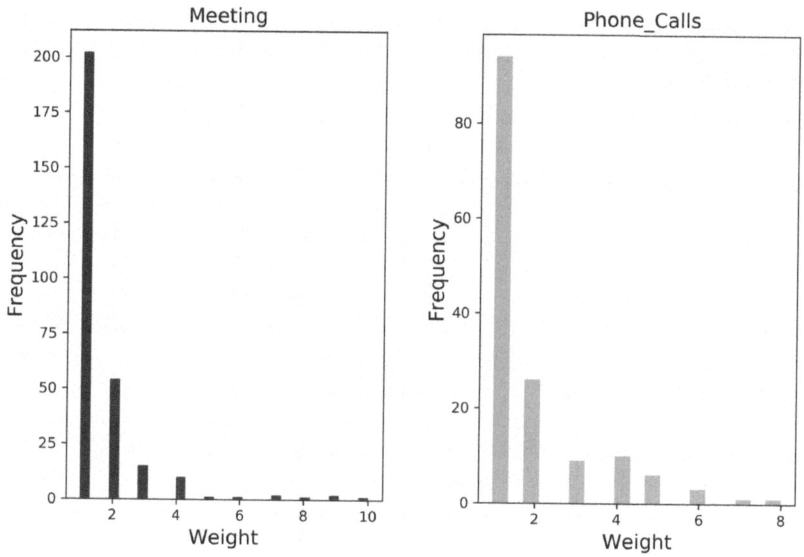

(a) MEETINGS (b) PHONE CALLS

Fig. 2 Weight distribution in the MEETINGS dataset (**a**), and the PHONE CALLS dataset (**b**). (Reproduced from Ficara et al. 2020)

Even through this analysis, as it was for the weights distribution, it emerged that the clan tries to minimize the risk of interceptions, especially to avoid exposing those mobsters who are hierarchically in a higher rank.

4.4 Degree Distribution Analysis

The Degree Distribution Analysis has been conducted in order to discover an appropriate artificial network that would virtually mirror the real-world criminal graphs topology under scrutiny. We previously analysed the weight distribution, but due to lack of libraries available for weighted graphs analysis,[2] we opted for a preliminary analysis on nodes degree distribution. To this end, in Fig. 4 we compared our real criminal networks against five artificial models: (i) the Random Network by Gilbert (G-ER), (ii) Watts-Strogatz (WS), (iii) its variant accordingly with Newmann [33] (N-WS), and (iv) two different configurations of the Barabási-Albert (BA) model in terms of links added at each step; BA2 with $m = 2$, and BA3 with $m = 3$. Tables 2 and 3 summarize the number of edges and average degree obtained with the configurations above described in *Phone Calls* and *Meetings* graphs, respectively.

[2] https://networkx.github.io/documentation/networkx-1.9/reference/generators.html.

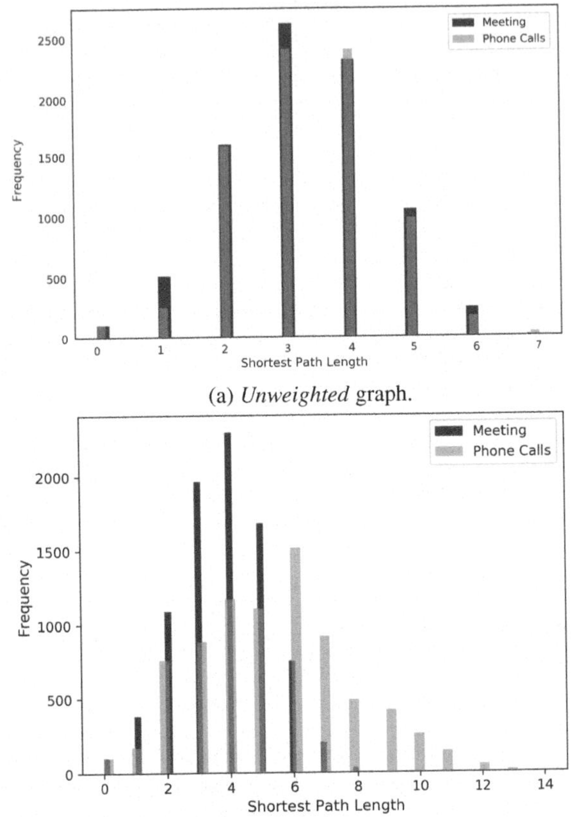

(a) *Unweighted* graph.

(b) *Weighted* graph.

Fig. 3 Distribution of shortest path lengths in MEETINGS and PHONE CALLS networks in Unweighted (**a**) and Weighted (**b**) graphs. (Reproduced from Ficara et al. 2020)

Note that all the results herein shown represent the average results obtained after 100 runs per each synthetic network.

We initially compared the real networks with the Erdős-Rényi (ER) topology, but this model did not allow to customize the number of links. Then, we opted for the Gilbert one, whereby both number of nodes n and links m are defined a priori.

In the WS model, we set $n, k = \frac{2m}{n}$ (that represents the number of nearest neighbors links per node), and the rewiring probability $p = 0.5$, with $p \in [0, 1]$. As previously asserted in Sect. 2.2, if $p = 1$, we turn into a Random Network. The main difference between WS and N-WS models is that in WS, number p is the probability of rewiring each edge; whereas in N-WS, $p = 0.25$ is the probability of adding a new edge for each edge. Indeed, if $p = 1$, then number of edges is doubled.

The actual graphs degree distributions act very differently from one another. In particular, the fluctuations in *Meetings* are justified by the fact that face-to-face encounters have been observed not only between couple of suspects, but also among groups of more than two people at the same time. On the other hand, phone calls have only been considered between individual suspects. The analysis suggests that through

Fig. 4 Degree Distribution in the MEETINGS dataset (**a**), and the PHONE CALLS datasets (**b**). Circles give the actual datasets values. G-ER is the Random Network proposed by Gilbert. WS is the Watts-Strogatz network. N-WS is the Newmann variation of WS. BA2 and BA3 are the Barabási-Albert models with $m = 2$ and $m = 3$, respectively

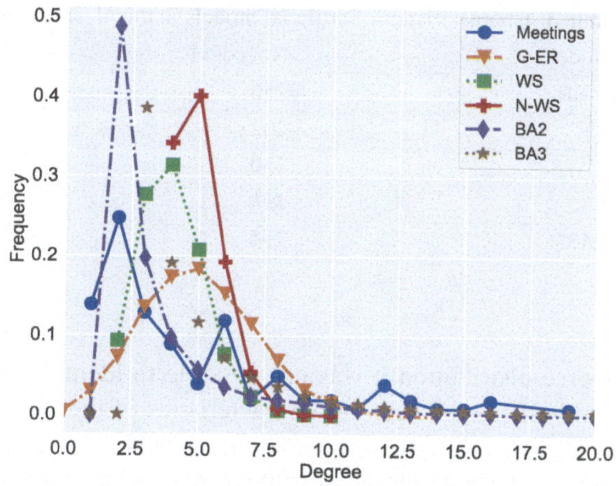

(a) MEETINGS

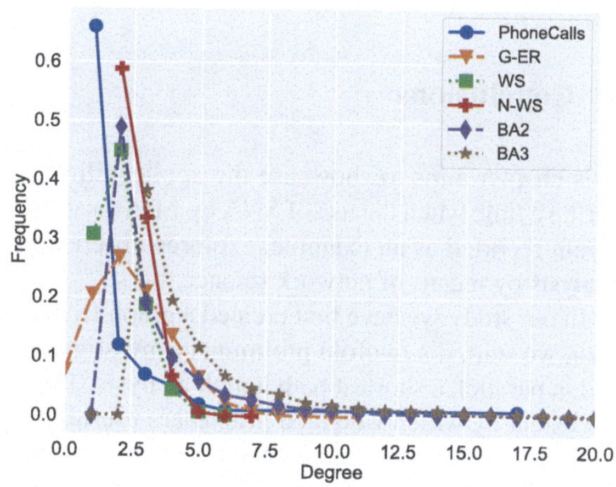

(b) PHONE CALLS

Table 2 Characteristics of artificial models in the *Phone Calls* network

Model	No. edges	Avg. degree
G-ER	124	2.48
WS	100	2.00
N-WS	123	2.46
BA2	196	3.92
BA3	291	5.82

Table 3 Characteristics of artificial models in the *Meetings* network

Model	No. edges	Avg. degree
G-ER	256	5.07
WS	202	4.00
N-WS	250	4.95
BA2	198	3.92
BA3	294	5.82

degree distribution it was not possible to identify an appropriate artificial network that best fits the network characteristics of the two real-world datasets considered. This is mainly due to the size of the networks. In fact, artificial networks seem to work better with lager sizes; thus, are quite unstable in the first step of their creation. For example, the emergence of hubs in BA models cannot be highlighted because of the small size of the overall network obtained.

5 Conclusions

This chapter aims to showcase the applicability of Graph Theory in Criminology, during a time when the use of SNA by LEAs is growing substantially. The case study herein reported as an example, explores different approaches on criminal networks analysis by means of network science tools.

In our study we have first created a graph from data extracted from juridical acts; then, we started a twofold preliminary investigation: a weight distribution analysis, and in parallel, a shortest path length analysis. These have been conduced to identify the extent by which weighted graphs are useful in those small networks.

Thus, we conducted a comparative degree distribution analysis between our real-world networks and some models generated by popular artificial networks. The aim was to identify the appropriate synthetic network which could simulate criminal networks artificially, but in an effective manner. The strength of this idea is that we may also be able to understand the patterns followed by criminals to create their internal interconnections among affiliates. Our study has found that the network size is a limitation. Indeed, there are significant fluctuations and through degree distribution comparative analysis it was not possible to find an appropriate artificial network that accurately mirrors the two real criminal networks used in our tests.

To overcome this issue, in future studies we will investigate adjacency matrix structures for both real and synthetic networks to get insights into network topologies.

References

1. Linyuan, L., Zhou, T.: Link prediction in complex networks: a survey. Phys. A Stat. Mech. Appl. **390**(6), 1150–1170 (2011). https://doi.org/10.1016/j.physa.2010.11.027
2. Hasan, M.A., Zaki, M.J.: A survey of link prediction in social networks. In: Aggarwal, C. (ed.) Social Network Data Analytics. Springer, Boston, MA (2011)
3. Tassiulas, L., Katsaros, D., Basaras, P.: Detecting influential spreaders in complex, dynamic networks. Computer **46**(4), 24–29 (2013). https://doi.org/10.1109/MC.2013.75
4. Cavallaro, L., Costantini, S., De Meo, P., Liotta, A., Stilo, G.: Network connectivity under a probabilistic node failure model. In: ArXiv e-print (Jun 2020). arXiv: 2006.13551 [cs.SI]
5. Mocanu, D.C., Mocanu, E., Stone, P., Nguyen, P.H., Gibescu, M., Liotta, A.: Scalable training of artificial neural networks with adaptive sparse connectivity inspired by network science. Nat. Commun. **9**(1), 1–12 (2018)
6. Cavallaro, L., Bagdasar, O., De Meo, P., Fiumara, G., Liotta, A.: Network science strategies for accelerating the training of artificial neural networks. In: Numerical Computations: Theory and Algorithms NUMTA 2019, p. 169 (2019)
7. Cavallaro, L., Bagdasar, O., De Meo, P., Fiumara, G., Liotta, A.: Artificial neural networks training acceleration through network science strategies. In: Sergeyev, Y.D., Kvasov, D.E. (eds.) Numerical Computations: Theory and Algorithms, pp. 330–336. Springer International Publishing, Cham (2020)
8. Bellur, U., Kulkarni, R.: Improved matchmaking algorithm for semantic web services based on bipartite graph matching. In: IEEE International Conference on Web Services (ICWS 2007), Salt Lake City, UT, pp. 86-93 (2007)
9. Zhou, X., Zafarani, R.: Fake news: a survey of research, detection methods, and opportunities. In: ArXiv e-print (Dec 2018). arXiv:1812.00315 [cs.CL]
10. Oluwabunmi, O., Cosma, G., Liotta, A.: Clan-based cultural algorithm for feature selection. In: 2019 International Conference on Data Mining Workshops (ICDMW), pp. 465–472. IEEE (2019)
11. Ficara, A., Cavallaro, L., De Meo, P., Fiumara, G., Catanese, S., Bagdasar, O., Liotta A.: Social network analysis of sicilian mafia interconnections. In: Cherifi, H., Gaito, S., Mendes, J., Moro, E., Rocha, L. (eds.) Complex Networks and Their Applications VIII. COMPLEX NETWORKS 2019. Studies in Computational Intelligence, vol. 882, pp. 440–450. Springer, Cham (2020). https://doi.org/10.1007/978-3-030-36683-4_36
12. Cavallaro, L., Ficara, A., De Meo, P., Fiumara, G., Catanese, S., Bagdasar, O., Song, W., Liotta, A.: Disrupting resilient criminal networks through data analysis: the case of Sicilian Mafia. PLoS ONE. **15**(8), e0236476 (2020). https://doi.org/10.1371/journal.pone.0236476
13. Barabási, A.L., Pósfai, M.: Network Science. Cambridge University Press, Cambridge (2016). http://barabasi.com/networksciencebook/
14. Antoniou, Ioannis: E and Tsompa, ET : Statistical analysis of weighted networks. Discret. Dyn. Nat. Soc. **2008** (2008). https://doi.org/10.1155/2008/375452
15. Barthélemy, M., Barrat, A., Pastor-Satorras, R., Vespignani, A.: Characterization and modeling of weighted networks. Phys. A Stat. Mech. Appl. **346**(1–2), 34–43 (2005). https://doi.org/10.1016/j.physa.2004.08.047
16. Travers, J., Milgram, S.: The small world problem. Psychol. Today **1**(1), 61–67 (1967)
17. Milgram, S.: An experimental study of the small world problem. Sociometry **32**(4), 425–443. American Sociological Association (1969). https://doi.org/10.2307/2786545
18. Erdős, P., Rényi, A.: On random graphs I. Publicationes Mathematicae **6** 290–297 (1959)
19. Watts, D.J., Strogatz, S.H.: Collective dynamics of small-world networks. Nature **393**, 440–442 (1998)
20. Barabási, A.L., Albert, R.: Emergence of scaling in random networks. Science **286**, 509–512 (1999)
21. Chen H., Chung W., Xu J., Wang G., Qin Y., Chau M.: Crime data mining: a general framework and some examples. IEEE Comput. **37**, 50–56. IEEE (2004). https://doi.org/10.1109/MC.2004.1297301

22. Bahulkar, A., Szymanski, B.K., Baycik, N.O., Sharkey, T.C.: Community detection with edge augmentation in criminal networks. In: 2018 IEEE/ACM International Conference on Advances in Social Networks Analysis and Mining (ASONAM), pp. 1168–1175. IEEE (2018)
23. Berlusconi, G., Calderoni, F., Parolini, N., Verani, M., Piccardi, C.: Link prediction in criminal networks: a tool for criminal intelligence analysis. Public Library of Science. PLoS ONE. **11**(4), 1–21 (2016). https://doi.org/10.1371/journal.pone.0154244
24. Franchetti, L., Sonnino, S.: La Sicilia nel 1876. **1**. Barbèra G. (1877)
25. McGloin, J.M.: Policy and intervention considerations of a network analysis of street gangs. Criminol. Public Policy **4**(3), 607–635 (2005). https://doi.org/10.1111/j.1745-9133.2005.00306.x
26. Mastrobuoni, G., Patacchini, E.: Organized crime networks: an application of network analysis techniques to the American Mafia. Rev. Netw. Econ. **11**(3) (2012). https://doi.org/10.1515/1446-9022.1324
27. Krebs, V.: Mapping networks of terrorist cells. Connections **24**(3), 43–52. INSNA (2002)
28. Xu, J., Chen, H.: Criminal network analysis and visualization. Commun. ACM **48**(6), 100–107. ACM (2005). https://doi.org/10.1145/1064830.1064834
29. Calderoni, F., Morselli, C.: Inside criminal networks. Eur. J. Crim. Policy Res. **16**(1), 69–70 (2010). https://doi.org/10.1007/s10610-010-9118-7
30. Campana, P., Varese. F.: Listening to the wire: criteria and techniques for the quantitative analysis of phone intercepts. Trends Organ. Crime **15**(1), 13–30 (2012). https://doi.org/10.1007/s12117-011-9131-3
31. Cavallaro, L., Ficara, A., De Meo, P., Fiumara, G., Catanese, S., Bagdasar, O., Song, W., Liotta, A.: Criminal Network: The Sicilian Mafia. "Montagna Operation" (Version 0.0.1) [Data set]. Zenodo. (2020). https://doi.org/10.5281/zenodo.3938818
32. Castaldo, F. (ed.): Messina, arrestati il capo ed i sodali della "Famiglia mafiosa di Mistretta". In: Grandangolo, il giornale di Agrigento (Jan. 18th 2019). https://www.grandangoloagrigento.it/mafia/messina-arrestati-il-capo-ed-i-sodali-della-famiglia-mafiosa-di-mistretta
33. Newman, M.E.J., Watts, D.J.: Renormalization group analysis of the small-world network model. Phys. Lett. A **263**(4), 341–346 (1999). https://doi.org/10.1016/S0375-9601(99)00757-4

A Data Mining Approach for Indoor Navigation Systems in IoT Scenarios

Mahbubeh Sattarian, Javad Rezazadeh, Reza Farahbakhsh, and Omid Ameri Sianaki

Abstract Indoor positioning is an important aspect in internet of things which plays a crucial role in various scenarios. Meanwhile, depending on the scenario, Indoor Navigation Systems (INS) generates considerable amount of data which can be used in navigation as a data-driven approach. In this paper, a navigation method has been proposed based on data mining techniques which enables value added services for end-users in different IoT scenarios such as airports, shopping mall and hospitals. The proposed heuristic algorithm is based on combining the greedy and random forest algorithms. Toward that end, we have collected data from a real world scenario (a large hospital) and used them to our implementation. According to our results regarding passage of time and accumulated history in the central server, we were able to suggest better routes while the proposed method shows reduction in both traveled distance and elapsed time. It also improved routing by decreasing the number of turns and encounters with obstacles. Thus, the proposed method provides better solution as an intelligent indoor navigation.

Keywords The Internet of Things (IoT) · Indoor navigation · Localization · Machine learning · Random forest · Data generation

---

M. Sattarian
Tehran North Branch, Islamic Azad University, Tehran, Iran
e-mail: m.sattarian@outlook.com

J. Rezazadeh
KENT Institute Australia, Sydney, Australia
e-mail: rezazadeh@ieee.org

R. Farahbakhsh (✉)
Institut Mines-Télécom, Télécom SudParis, CNRS Lab UMR5157, Paris, France
e-mail: reza.farahbakhsh@it-sudparis.eu

O. Ameri Sianaki
Victoria University Business School, Flinders Campus, Melbourne, Australia
e-mail: omid.amerisianaki@vu.edu.au

G. Fortino et al. (eds.), *Data Science and Internet of Things*, Internet of Things,
https://doi.org/10.1007/978-3-030-67197-6_9

1 Introduction

Recent Internet of Things (IoT) advances have shown the importance of information technology in business more than ever before. Organizations need efficient data access to be leveraged in different applications. User information helps organizations to provide user services more appropriately, perform purposive marketing, and allocate resources optimally. Accordingly, the IoT is regarded as a high-speed technology, mainly based on the idea of connecting every piece of physical equipment to each other [1]. Application scenarios of IoT cover a wide spectrum including health, smart city, building management, which receive information from the environment and take appropriate action based on environmental conditions. A navigation system plays a major role as the procurement and supply chain management in the IoT applications. Position estimation is of great importance in both indoor and outdoor environments for industries and universities. Obviously, the successful application of the Global Positioning System has enabled people to travel freely around the world [2].

Although, GPS cannot be used indoors, the received Wi-Fi signal strength is measured indoors for routing. The goal of a smart building might be to record human activities in indoor environments. Accurate positioning is necessary for commuting routes in complexes with many floors such as exhibitions, airports, hospitals, railway stations, shopping centers, and industrial buildings. Only 50% of the solution comes from positioning an individual or a device indoors. If service providers wish to implement an indoor navigation system for certain purposes, they need accurate drawings of indoor environments. Data mining can be described as a process in which data are analyzed to find patterns, procedures and discover latent relationships. As a result, experts can predict the expected information. Data mining focuses on an idea to show a better path providing certain advantages, i.e. time and cost. Data mining can be used in various indoor navigation systems to estimate patterns and procedures. It is a general solution to the attainment of a competitive advantage. In the past, it was only possible to evaluate what INS devices or user did. Nowadays, data mining techniques can be employed to predict what actions they will take. With the use of data mining techniques in INS, better advertising and marketing decisions can be made. The main contributions of this study are summarized as follows:

- A dataset generating algorithm was implemented which utilizes user arrivals and motions from a real environment and provides a set of data on the user travel history. Eventually, an indoor navigation system is modeled efficiently.
- Designing a navigation model based on data of a real world scenario which is precisely adapted from the emergency ward of a large hospital. After data classification, a heuristic algorithm was developed by the combination of greedy and random forest algorithms, which improves routing.
- By analyzing the accumulated history data in the central server, previous experiences were used to suggest better routes, our approach shows improvement in terms of traveled distance and elapsed time by decreasing the number of turns and encounters with obstacles.

2 Related Work

Internet of things (IoT) is a new paradigm consisted of billions tiny sensors; these sensors establish a network, which communicate with each other to help solve real time problems. Besides, IoT applications including supply change management and logistics have been significantly under the influence of positioning systems [3–6]. Nowadays, life has become easier and comfortable due to widespread use of smart devices. Researchers have predicted that billions of embedded devices, termed as Internet of things (IoT), will dominate communications by 2020 in comparison to recent human interactions [7]. The location of things can be used to provide a wide range of location based services in IoT paradigm [8]. Therefore it will need to be able to track and navigate objects in IoT world [9, 10].

Recently, smart-navigation based IoT has caught researchers' attention in both academia and industry. Due to drawbacks of traditional methods, smart indoor navigation systems have been introduced [11]. IPS is the terms used to explain the process of the relative position of the other static or dynamic objects located in the designate area [12]. Researchers have attempted to the indoor positioning technique based on the indoor communication infrastructures for different applications of IoT; According to 2018 estimations, the domestic market value of these techniques are 5 billion in 2018 [13]. GPS is a good option for outdoor positioning, but its results for indoor positioning are not desirable; since there is a line of sight (LOS) transmission path in GPS, it causes shadowing, fading, and delay distortion in indoor environment [14]. Recently, various indoor positioning methods have been introduced using different radio and non-radio based technologies. RFID, WLAN [15], Zig bee, UWB, bluetooth [16] and cellular are radio based technologies and acoustic and thermal sensors are among non-radio based ones [17].

In indoor navigation systems, it is assumed that certain disturbances outside buildings could be encountered with an internal magnetic field. Nevertheless, it should be noted that the magnetic field inside buildings is quite constant. Determining the distance between the receiver and transceiver is usually done by RF-based systems using RSS. Wi-Fi technology is currently used extensively for localization purposes. Since LED luminaries are used extensively in buildings, visible light communication systems are one of the best options for localizing and navigating purposes in indoor environments [18]. Visible light communication technology (VLC) and its facilities has gain enormous interest for indoor positioning in recent years.

In [19], a hybrid positioning method has been introduced; they used the fusion of power line and visible light communication systems that could be used for indoor navigation in many location based health services such as first aid treatment and so on. Zhang et al. [17] highlight the precision of VLC-based IPS; they have analyzes it in simulation environments. The challenges that lay ahead are dealt with in this study too. Currently, navigation in complex environments such as exhibitions, hospitals, shopping malls, airports, train stations, office and university buildings, hotel resorts etc. have caught the researchers' attention. Undoubtedly, navigation could be influenced by IoT solutions. Due to rampant use of mobile devices and their ever increasing capabilities, they have been considered for location based services.

Apple Inc. has recently introduced iBeacon technology, which is based on Bluetooth Low Power (BLE) and much more energy-efficient than WiFi and classic Bluetooth technologies [20]. In this respect, a comprehensive survey is found in [21]. One of the smart phone based navigation system is NavCog [22], it applies Blue-tooth low energy (BLE) beacons and K-nearest neighbor (KNN) algorithm, which outperforms GPS and Wi-Fi based methods. LowViz [23] is an indoor navigation application; it uses various technologies such as Wi-Fi and Blue-tooth low-energy beacons and inertial sensors to render precise results. A vision-guided navigation system was introduced by Zheng et al. [24], the users can use this system to launch their own indoor navigation services feasibly. The users can use the application on their smart devices and search through pathway images and indicating turns, etc. An indoor positioning method and system has been introduced by Wang et al. [25] based on I Beacons used for signal emission. Smart phones receive the emitted signal using RSSI (Received Signal Strength Indication), which is applied for signal obtainment. Thus, trilateration method can be used to pinpoint the position [26].

Indoor navigation by definition is helping people to find their location, detect, and avoid obstacles; indoor navigation is the field in which the problems of locating a person to explore several domains ranging from emergency response to enhanced marketing strategies in micro indoor environments are solved [12]. Technological advanced have always tried to help people enhance their life quality, especially disabled individuals or people with limited abilities. Sight is the most important sense in humans. The only way for a sightless individual to interact with the world is through external aids or devices. Recently, various researches have focused on indoor navigation systems for blind people [27]. BlindeDroid has been introduced in [28], this application helps blind people using their location data collected by beacons placed in strategic locations inside a building. With this information, the application guides the user's navigation, answers his/her questions and provides them with useful information concerning places, products, and services near the user.

An autonomous indoor navigation system for blind people, called Blind Museum Tourer, was introduced by Meliones et al. [29]. It was designed to help blind people on self-guided tours in museums. The application server receives the current blind position and positions of other blind along the navigation path. In case of a likely collision, the application swiftly notifies the user in advance. In [29], a modern navigation system is introduced for pedestrians. The system has been integrated into an application interacting with a small embedded system that reads simple user controls, tracks pedestrian mobility using GPS, and detects near-field obstacles and traffic light status. By emerging wireless-source and strong-signal dependent IoT, more research has been conducted on indoor navigation [20]. It has been attempted to use different navigation technologies by IoT for indoor navigation [30]. A local database or a central database is usually the sources for obtaining information by a navigation system; this information is received through wireless connection. Studies on data mining user information in an indoor navigation system are not adequately done. In our previous survey [31], the most current research on data mining techniques for indoor navigation systems using IoT has been presented; it was concluded that more accurate decisions could be made by the navigation systems using data mining techniques. Zhu et al. [32] have focused on deep reinforcement learning techniques for visual navigation. Using a large number of training samples made their proposed method faster and more generalized than other techniques.

Inspired by the advances in cognitive science, artificial intelligence, computer vision and mobile computing technologies, a virtual tool is proposed for indoor navigation based on autonomous visual perception. A Bayesian network model for route planning is provided according to the initial position and destination [33]. The results show that their method has improved accuracy by over 13% compared to those based on the hidden Markov model. Multiple assisted indoor navigation and tracking (MINT) is a localization method, by which the traditional decentralized method was employed along with a central node to measure the distances between connected nodes. Osama et al. [34] have proposed a combination of long-term capacitive sensors and machine learning classifications for indoor environments to compare the localized performance of ML algorithms and their changes in relation to the required training set and resource algorithm for both the training set and the results. They examined the performance of many ML classification algorithms and found out that the random forest algorithm has the best performance among them. Several papers relied on deep learning structure based localization such as [35] which their proposed method achieves state of the art performance on positioning with Wi-Fi signals. Although machine-learning algorithms have been widely studied, little attention has been paid to the field of data mining in indoor positioning systems.

After extensive investigation of existing state of the arts and based on authors knowledge, no research has been conducted by leveraging data mining of user information for indoor navigation. Therefore, based on extensive research carried out in the area of indoor navigation systems and data mining, it was decided to change the approach and focus on navigating rather than positioning. The research in the field of indoor positioning is based on the fact that the internal magnetic field creates disturbances from the outside of buildings, while it is relatively constant inside the building. This approach lead to the idea of developing an indoor navigation system based on data collected from different circumstances by using machine learning techniques to suggest a better route in which real-time interactions between users, objects, and indoor positioning systems are recorded in applications.

3 The Proposed Method

At first, preparation complexities and absence of available dataset, which truly records user traffic, motivated us to develop a dynamic simulator with a web-based user interface for data generation. More specifically, we developed our simulator to exert paths, obstacles, and patterns in order to generate datasets and analyze our algorithms. Then, owing to necessity of a space for data storage and analysis, a database was created. According to previous studies on the use of intelligent sen-

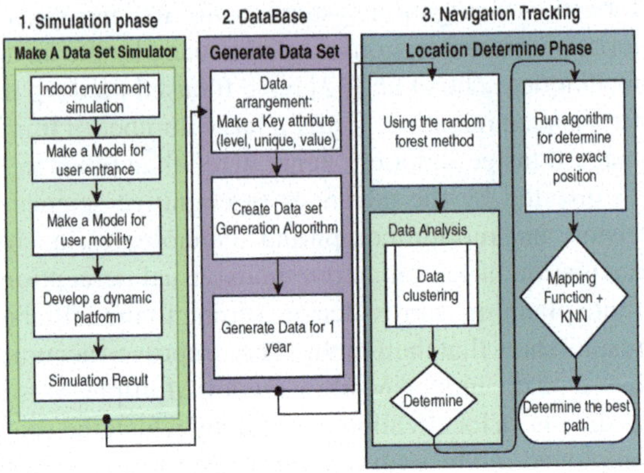

Fig. 1 Schematic diagram of study steps

sors and sensor data for indoor navigation (reviewed in Sect. 2), certain decisions were taken to change approaches and focus on navigation instead of positioning. The studies on indoor navigations are based on the idea that internal magnetic field causes disturbance outdoors and remains relatively constant indoors. This approach laid the foundation for an indoor navigation system based on the data collected from different positions. Finally, the combination of random forest and greedy algorithms were utilized for data analysis. This system is also equipped with machine learning techniques to improve route suggestion. Figure 1 shows a schematic diagram depicting the steps of the study. In the first step, different simulators were evaluated to map real indoor places and develop an indoors model for user movements. Finally, the PHP programming language was used to create a simulator generating data based on the proposed model. In the second step, a MySQL database with the required records was created and connected to the simulator. After implementing the project, the necessary datasets were generated over one month and a year respectively. In the third step, the random forest algorithm (known as a machine learning algorithm) was applied on the data generated at the previous step. According to the results, the proposed method reduced the traveled distance and elapsed time. It also improved routing by decreasing the number of turns and encounters with obstacles. As the history records were expanded in the central server over time, better routes will be suggested using the previous experiences.

3.1 Simulating an Indoor Environment

The PHP programming language was employed to simulate (10×10 m) space mapping at the emergency ward of a large hospital. The environment was divided into 100 equal segments (1 m in both width and length), and we assumed five iBeacon sensors were embedded at proper distances to establish communication with user mobile phones. Figure 2 shows the details of modeled environment and Fig. 3

Fig. 2 The modeled indoor environment

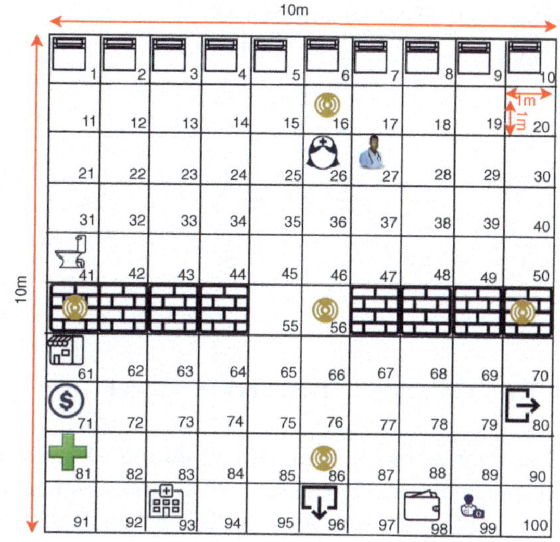

Fig. 3 Details on recorded positions

	Emergency beds are located at locations 1 to 10.
	Sensors are located at locations 16, 51,56,60,86.
	The nurse's room is located at location 26.
	The doctor's office is located at location 27.
	The toilet is located at location 41.
	Entrance and exit doors are located at locations 80 and 96
	The store is located at location 61.
	The ATM is located at location 71.
	The drugstore is located at location 81.
	The triage section is located at location 93.
	Emergency's cache register is located at location 98.
	Emergency release is located at location 99.

presents the aforementioned positions accurately. The following rules were observed in simulation of the emergency ward in the hospital:

- The positions of highlighted components were fixed with different priorities on the grid which is precisely modeled from the real environment.
- Upon arrival, every user complies with a certain priority, for instance, a user must visit the triage ward after arrival.

Table 1 Detail on user arrival model

Period	Percent (%)	Number of users
12–8	30	300
16–12	20	200
20–16	25	250
24–20	15	150
4–24	5	50
8–4	5	50

3.1.1 Creating a User Arrival Model

The mean arrival time of patients during 24 h was obtained at the emergency ward of the hospital to design a user arrival model and generate the target dataset. According to the obtained result files, 1000 individuals visited the emergency ward at different times of a day on average. Table 1 shows the mean arrival time during a day in details. Then, a Gaussian distribution function was used along with a series of random functions to create 1000 users in the simulator. Distribution function were employed to enter the users at specific nonrandom locations during a 24 h period.

3.1.2 Creating a User Motion Model

The greedy algorithm is known to be one of the best existing algorithms for routing problems. In this study, the user movements were not random, but intelligently controlled by the greedy algorithm. As previously mentioned the considered positions were also nonrandom and specified accurately. Based on the greedy algorithm, the users move towards their randomly selected targets, the priorities of which may vary depending on the location. Users may randomly move between positions and wait within a time interval (specified with a maximum and minimum) to receive certain services, after which they leave the positions. According to the motion system, users comply with the greedy algorithm and select the shortest route as they move towards a target without encountering any obstacles. However, they may also use the wall follower, right-handed, or left-handed algorithms when they encounter congestions or obstacles to change routes. In other words, the user may move along the wall to bypass an obstacle. This method is not entirely accurate because an artificial intelligence user acts like a real human who has entered an unknown place for the first time and may find a route by trial and error. In fact, the users may not comply with the triangle inequality to select the shortest route given that they are not able to foresee the traffic in their path. Hence, errors are inevitable in this approach and for instance, a 20-steps length path may be traversed in 30-steps by the user. Following rules have been set to the user motion model:

1. On average, individuals are added randomly at different intervals during the 24 h of a day.
2. Weighted targets are selected for users randomly. In other words, every position is given a specific weight in the model mapped from the emergency ward of the hospital. Moreover, a counter counts the number of times a position is randomly selected and approached by users.
3. Users' next positions is selected by the greedy algorithm. In other words, after arrival of patients, the greedy algorithm decides their next move towards the destinations, which were randomly selected based on the priorities registered for every position.
4. When users encounter walls, they use the wall follower algorithm to find a route.
5. Users motions are recorded in different locations every second during the 24 h of day from their time of arrival until their departure. The motions are stored in a MySQL table.
6. Certain time intervals during which no users are present in a given position are left out, as they are considered null.
7. The database contains as many rows as the number of users entering the model space during the 24 h of day.

3.2 Designing a Dynamic Platform

To generate the required dataset, a platform was designed which may also be used in other studies. It can be employed to simulate every new environment dynamically by defining positions, capacities, and attributes to generate the necessary dataset. Finally, certain algorithms can be used in the same simulator to analyze and evaluate outputs dynamically. The codes are freely available in GitLab and can be easily provided to other scholars for research purposes. As shown in Fig. 4, the map and positions can be adjusted. A fixed framework was laid out for dataset generation. The maximum and minimum capacities of positions should be specified along with their priorities. Although the total number of users registered in the system during one day is among the constant entries, their registration time is random.

3.3 Components of the Simulator

Existing navigation process models (i.e., [36–38]) can be considered as a machine with finite state and fixed logic conditions for state transfer that cannot count non-constant received factors. Since space recognition involves both cognitive and conceptual processes, [39] the model of the navigation process requires the integration of perception and recognition in real scenarios. We propose an algorithm within a dynamic process for navigating and managing uncertainty. The proposed algorithm considers non-constant factors as a secondary factor and has indirect impact on the

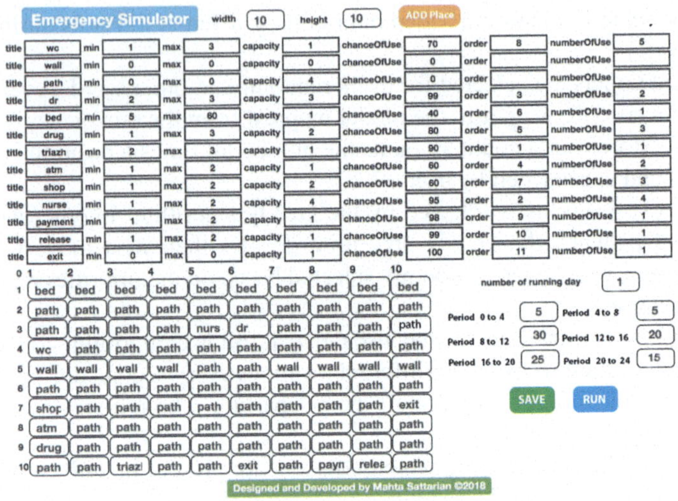

Fig. 4 Dynamically designating the capacities and constraints of each position (UI view)

Table 2 The configurations set for simulation

Parameter	Value
Simulated area	10*10
Number of obstacles	8
Number of runs	10 times
Number of sensors	5
Real time data	1 Day
Training data duration	1 year

proposed path. For further investigation, we created our model by simulating the emergency ward of a real hospital in which PHP 7.0.0 was used for developing the platform, and MySQL was used as the database. In user interface, j Query and Bootstrap were employed. Moreover, an open-source Java Script framework for building user interfaces named VUE.JS was used as UI platform. Then, Chart.JS was used to generate simple HTML-based Java Script charts. Table 2 indicates the details on configurations for the environment simulation and program execution.

4 Creating a Database to Store the History of User Motions

In this section, we introduce our strategy on a key part of this study, which generates the required dataset by means of a new simulator designed by ourselves. Towards that end, MySQL was connected to PHP in order to store the necessary records of user's history motion. In other words, 1000 users are registered in the system at each execution. The motion information is recorded for every route in the database which includes a user ID ($User - ID$), registration time ($time$), departure time (end), starting position ($start - pos$), leaving positions ($end - pos$), elapsed time ($duration$), the number of steps ($length$), and traveled distance ($path$). The recorded real world data adaptation are used as training data.

4.1 Dataset Generation Algorithm

The program was entirely developed based on the object-oriented programming paradigm. Patients and locations were labeled as Person and Location, respectively. The considered classes are as follows:

- Person: This class includes certain variables such as user IDs, arrival time, departure times, traveled routes with respect to time, current location, waiting time, and visited places.
- Location: This class includes only the coordinate IDs and the ID of those users present at these coordinates.
- Place: This class includes an array of locations, location types, the minimum and maximum time required for certain tasks, location capacity, probability of using a locations, the number of visits, and visiting priorities.

4.2 Components of Dataset Generation Algorithm

We adjust the program clock in seconds. At each clock, emergency attendants and newcomers will be given permission to move from one location based on the priority and arrangement of available places. The priorities and probability of positions are given in Table 3.

At any moment, there are nine possibilities during motion: standing in the current position or moving towards the other eight points around. These nine point are first sorted out in order of their distance from the destination. Then, the route with further capacity is selected. After arriving at the destination, user waits for a random duration ranging between the minimum and maximum waiting thresholds. In case another patient is already present at the considered location after reaching the destination, user waits for the previous patient to leave and then expects to receive services. Along the way, if there are a number of people on the line, their objectives are compared to that of the user. In case of different objectives, the user passes through the line. In other words, although each route can take four individuals, they are only lined up if they share the same objective, which is similar to a real-world scenario in the sense that people can cross waiting lines if they are not planning to receive any services from them. Another problem is the presence of walls. Upon encountering wall, users

Table 3 The priority and order of locations in each clock

Location	Admission capacity	Patients referral probability (%)	Min and max required time
Triage	1 person, each person maximum 1 time	90	Between 2 and 3 min
Nursing	4 person, each person maximum 4 times	95	Between 1 and 2 min
Doctor's office	3 person, each person maximum 2 times	99	Between 2 and 3 min
ATM	1 person, each person maximum 2 times	60	Between 1 and 2 min
Pharmacy	2 person, each person maximum 3 times	80	Between 1 and 3 min
BED	1 person, each person maximum 1 time	40	Between 5 and 60 min
Store	2 person, each person maximum 3 times	50	Between 1 and 2 min
WC	1 person, each person maximum 5 times	70	Between 1 and 3 min
Cache desk	1 person, each person maximum 1 time	98	Between 1 and 2 min
Release	1 person, each person maximum 1 time	99	Between 1 and 2 min

employ the TRIMAX algorithm to move around the walls. After receiving services from all the places, which were expected to be used with a probability of 80% or higher, patients will finally move towards the exit and then leave the emergency ward. In every execution of the program, all of the user motions are recorded in an array stored in MySQL.

5 Using Machine Learning Algorithms

Accurate indoor positioning is very important in many large and small scale IoT applications such as health monitoring and assisted living. Machine learning classifiers can effectively reduce the changeability of sensor data and the noises caused by environmental conditions. In this study, the empirical data were obtained by simulating an indoor localization system based on iBeacon sensors in a (10 × 10) m room. The machine learning results were analyzed from a general perspective; in addition, performance of localization algorithms, changes with respect to the training dataset, and the necessary resources were compared for both training and testing Datasets. The results indicate a large variance among most accurate algorithms. The presence of noise or environmental condition variables such as temperature, humidity, and brightness affects the accuracy of sensor data. Therefore, the raw data usually require considerable processing in order to achieve the localization accuracy required by programs. Regarding the data processing techniques, machine learning algorithms are the most promising ones. However, they can be significantly different in terms of performance (for instance inference, required training, and computational complexity). In this study, the random forest algorithm was used. The results indicate that these algorithms can produce good result on the selection of the best route. Moreover, the effects of size training dataset on the localization results were analyzed for this algorithm.

5.1 Using the Random Forest Algorithm

Random forest algorithm is one of the new methods in the classification of paths. The use of this algorithm is increasing due to satisfactory results, high speeds, and low parameters for adjustment in the classification [34]. Random forests are one of the special models in ensemble learning method. Different data mining techniques such as the logistic model, neural networks, decision trees, and support vector machines can be used to obtain various results. Combining these algorithms lead to achieving higher performance and lower error rate compared to using them individually. As a result, instead of making a single model, several models were created and their average results were considered. For this purpose, the ensemble learning method was selected to achieve the best model. Two popular methods of bagging and boosting were used for ensemble learning. In the former method, a large number of unstable models generate a stable model, so that a reliable prediction with a smaller variance is achieved in the new model. In this method, K bootstrap samples are extracted from the D dataset. Then a classification model is created for every bootstrap sample, generating a total of K models. The output of final classification is used for decision-making, and the mean regression output is selected as the result. It should be noted that the variance is reduced in mixed prediction, allowing us to reduce the unavoidable errors if the independent samples are available. The bootstrap sample comprised only 63% of the total samples, and the other 37% were not present at all. Thus, they differed from others. Finally, errors are eliminated in the independent models, and a better ensemble model is obtained with higher accuracy. Random forests are considered a better mixed learning method for regression classification. They work based on a structure consisting of numerous decision trees over training time and classification. Random forests are suitable for decision trees, which experience over fitting when trained by the training set. Among decision support methods, decisions trees and diagrams offer the following advantages:

1. Simple concept: Everyone can learn to work with decision trees through little study.
2. Working with large and complicated data: Despite simplicity, the decision tree can easily work with complicated data and make decisions based on them.

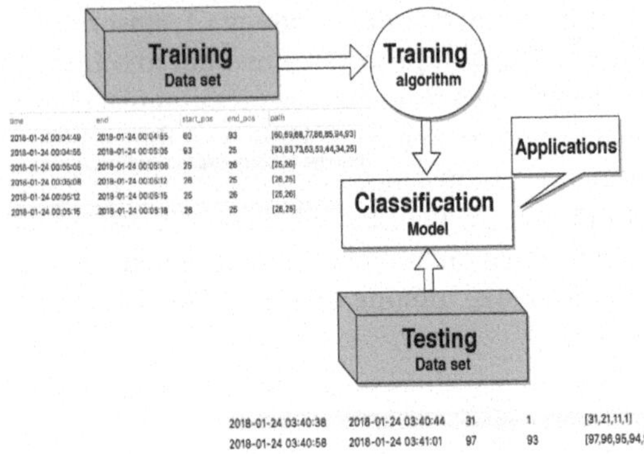

Fig. 5 Data classification

3. Re usability: If the decision tree is made for an issue, various examples of that problem can be calculated with that decision tree.
4. Combination with other methods: The result of a decision tree can be combined with other decision-making techniques to obtain better results.

5.2 *Data Classification*

According to Fig. 5, the data were classified as history and live data in the proposed method. Former is recorded in the system during previous days while the latter include current status of positions, accounting for 40% of data. The program was executed ten times with different percentages allocated to the history and live data. Finally, we considered 60% of the data history and 40% of the live data to suggest the route efficiently. Figure 6 indicates the improvement in algorithm results by allocating 60% of data to the history data. Following formula was used in the algorithms:

$$((\sum_{i=1}^{\infty} p_i) + Weight_h) * ((\sum p_L) + Weight_L)$$

In this formula, a classification is carried out by obtaining the mean of features such as distance, duration, and traffic level based on the source and destination of users. Finally, a weight ($Weight_h$) is assigned to the history data, and another weight ($Weight_L$) is given to the live users who traveled these paths. Furthermore, 60% of the decision for path selection was inferred from the recorded history and the remained was based on the greedy algorithm and the most suitable paths under the current conditions. This mechanism is similar to that of ant colony optimization algorithm.

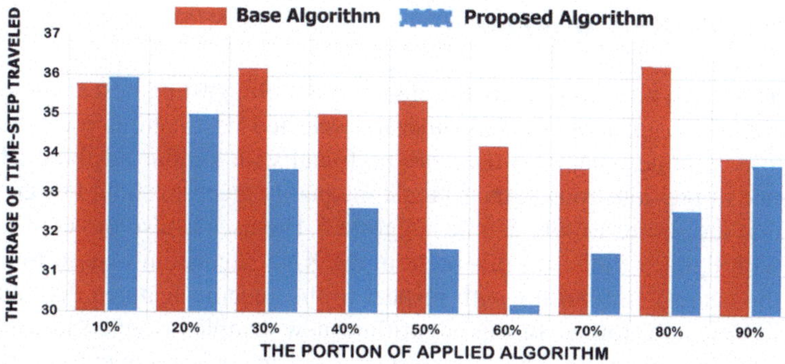

Fig. 6 Allocating 60% of data to the history

5.3 The Proposed Algorithm for Better Routes

The greedy algorithm was used to compute live data, and the history data were computed by employing the random forest algorithm, searching in the history recorded in the database, creating decision trees, and obtaining their mean. Based on the analysis of previous data, a route was suggested. Finally, the following formula was used to introduce the best route. Given the simulation conditions in this study, time and distance should be of the same priority. However, one might be more preferable compared to the other one in real implementation conditions at a hospital where certain condition may exist.

$(L + D) * \sqrt{1 + W^2} + B.$

Therefore, it was important to keep the length (L) and duration (D) variables at the same priority. Since no assumptions were made on the climatic conditions and urgencies, the effective factors were considered the presence of walls (W) and inaccessible positions due to saturation and other factors (B).

5.3.1 Mapping Function

In this step, a record named map was created in the database. Map is a function defining the factor weights. This function maps a set of parameters such as arrival time, source, destination, and traffic onto a number. Then the random forest algorithm was used to obtain a mean and suggest the best route to the user. Weighting factors include traffic, the number of steps, and elapsed time on route. As mentioned above, these factors were summed up into a single mean value to evaluate the information of present user in the system for 86400 s who are capable of making nine decisions in each moment.

5.3.2 Using the KNN Algorithm

It is assumed that a new user enters the environment. Suggesting best route to reach the destination is our goal. Arrival time and selected destinations are taken into account to determine whether a user can be found with similar attributes. Due to the presence of previous routes, the triangle inequality theorem should be complied with, so that the shortest route can be suggested. The user is also directed towards the path with lightest traffic. In this section, the KNN algorithm is used to suggest three of the nearest similarities with respect to the specific features defined, even if no similarities are found. In this algorithm, new samples (logged-in users) are compared to all previous instances (historical data) and are assigned to each of the earlier examples that are closer to those samples.

5.3.3 Computing a Better Route

A search is conducted in the designed model whenever a new user enters the system. Then, clustering is performed on the previous similar routes to assign their values. After that, decisions are made using the random forest algorithm based on a heuristic series in seconds. On the other words, the system learns training dataset to create a model. Then, an output is produced based on the considered parameters. Proposed algorithm considers certain parameter such as live traffic for searching important events in the database and forming a tree based on similar routes. Based on the steps taken and the map function, each tree is assigned to a number. Then, they will be sorted in database. Finally, the smallest number is selected as the best route. Greedy algorithm is used dynamically. So if the proposed routes encounter traffic, the algorithm re-selects the best route based on the calculations performed in different parts of the system. When a new user enters, the database is searched based on random forest algorithm. In case users of the same attributes are found within this threshold (time of arrival with a two-minute tolerance), the decision tree suggests a route based on the mapping function and the arrival day of the user. Proposed sets of trees are sorted in order of certain attributes such as length, duration, and traffic in the random forest algorithm. Then, the important factors are taken into account to vote on the best route.

6 Performance Evaluation

This study benefited from CRISP-DM, a standard data mining technique. It was introduced in late 1996 by three large corporations: Daimler-Chrysler (Benz), SPSS, and NCR [40]. This method proposes a process model for data mining to review the life cycle of every project. Such a life cycle consists of six stages: Understanding the business issue, perceiving data, preparing data, modeling, evaluating results, and applying model.

- Perceiving the business plan: in this step, the research goals are addressed from a business perspective. Given the development of IoT services and increased number of users, it is very important to identify users' requests and interests. The aim of this study is providing users with better routes obtained from a mixed process involving machine learning and analysis of different algorithms.
- Perceiving Data: after distinguishing the research goals and its relevance to business, the PHP programming language was used to store data in MySQL in addition to determining the appropriate criteria and simulating the human motion system and indoor environments. All of the necessary data were extracted automatically through different processes.
- Preparing Data: in this step, data cleansing was performed by taking certain actions such as eliminating incomplete data, unusable outliers, and missing values. Moreover, the extracted data can be integrated.
- Modeling: in this step, the greedy and wall-follower algorithms were used to design a model for user arrival and motion. Then, machine-learning algorithms were employed along with the generated training dataset to suggest better routes to users.
- Evaluation: According to the results of previous step and in comparison with outcomes in the absence of the proposed algorithm, the suggested model was evaluated.
- Applying Model: Creating the model does not mean the end of project. Even if the model intends to improve knowledge based on the given data, the results should be organized and presented, so that users can benefit from it.

Depending on the requirements of learning phase, a model can be used as easily as generating a report or as difficult as creating a repeatable data mining process. In [41] authors introduced a learning and inference model named LA-CWSN (learning automata-based cognitive wireless sensor network) for CWSNs. They employed the learning automata approach to add perception to general network stack protocols. LA is allocated to the important networks protocol parameters. Every automation has a finite set of possible values for the relevant parameters, which attempt to improve the network efficiency through learning. Given the applied nature of this study and the research findings can be used for simulations, data generation, in different areas related to the Internet of Things.

6.1 Selecting Sample Routes

Figure 7 shows the dispersion of all the routes defined in the system, which can be used by the users and are within the repetition range of 1–1400 times per day in various situations. These routes are defined based on the items and interpret priorities. The present study analyzes and evaluates the average of most frequent routes in the range of 100 to 20 based on Fig. 8. According to similar studies, the best and busiest routes were selected for evaluation after fifty times of running program. For example,

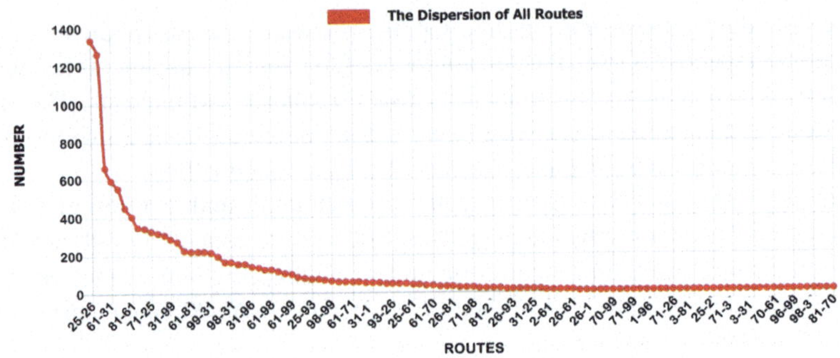

Fig. 7 Dispersion of all the routes

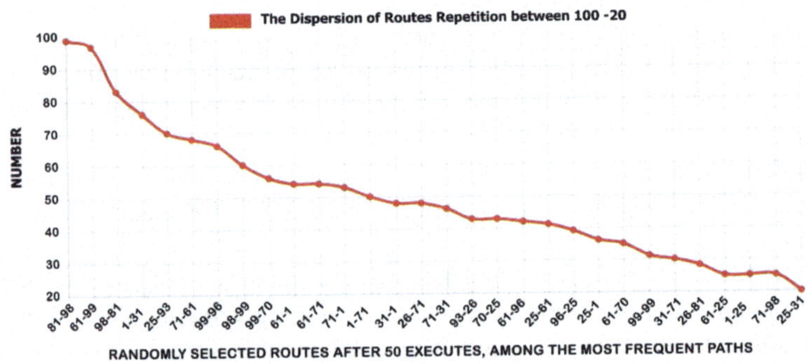

Fig. 8 The most frequent path (81–98), the least frequent path (25–31)

route 25-31, which is the route taken from the doctor's office to the restrooms, is the least frequented one, while route 81-98, which is the route from ATM to emergency rooms cash counter, is the most frequented one.

6.2 *The Effect of Algorithm on the Proposed Formula*

The x-axis in Fig. 9 chart includes start and end points of routes which are selected based on the priorities registered in the system, they are selected dynamically based on the specified range statements in the database. Y-axis is created based on the mapping functions (steps, time, and crowdedness) of the proposed route. Each time the program runs, a chart for the historical data (the blue chart) that has been stored in the database is mapped then the proposed algorithm is applied on it. At same time, a red chart is also mapped without applying the proposed algorithm. A decrease in the values of y-axis indicates the improvement caused by the proposed algorithm. As shown in Fig. 9, the training real world data adaptation had been stored for only a few days, after the application of proposed algorithm, only a 25% improvement was

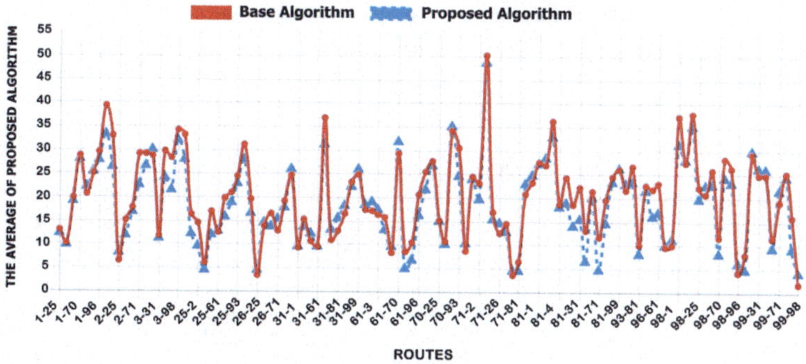

Fig. 9 The effect of proposed algorithm

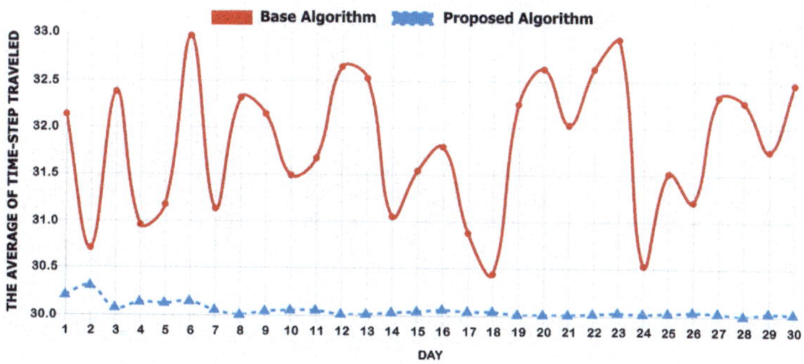

Fig. 10 Time-step traveled over one month

observed compared with the base algorithm. However, with increasing training data and storing them in the database, improvement increased significantly, as is evident from Figs. 10 and 11, which show a period of one month and a year, respectively. Based on the performance of learning algorithms in comparison with the base algorithm, there is no observable pattern in the red chart. That is, the obtained mean values are all stochastic and based on users' movement structures. However, as time passes, the mean values in the blue chart decrease and there is an observable improvement in the algorithm in following months, with decreased dispersion and more uniform results.

6.3 The Effect of Algorithm with Respect to the Time and Step Factors

In Figs. 12 and 13, the effect of algorithm is evaluated separately with respect to the time and step factors. In Fig. 12, which is time-based, routes 25-2 or 81-93 are the strength points of the proposed algorithm. There are low differences at some points

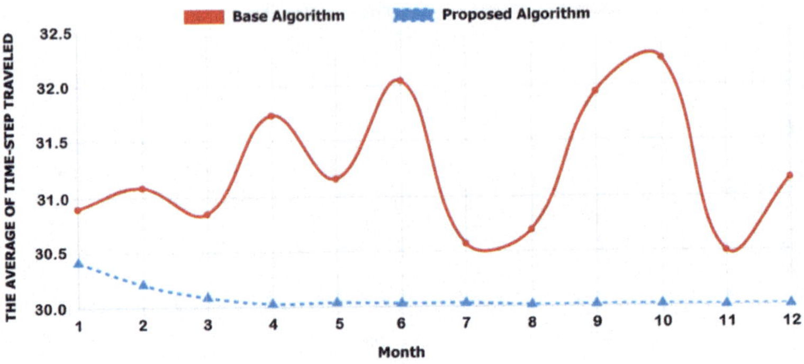

Fig. 11 Time-step traveled over one year

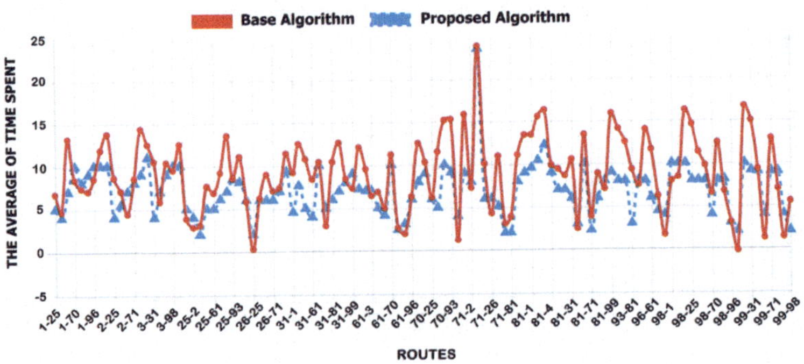

Fig. 12 Time spent

due to the absence of more optimal routes, hence the lack of learning capability. As observed, the two figures have identical results in some routes (e.g. 96-1 and 25-2) because of a direct, identical route in that region, which is not challenging or highly iterative. In some routes such as 31-61, there is a high difference between the two figures, indicating high iteration at the respective route, which leads to learning and generating optimal step-based routes. In the events of high congestion, the time is an effective factor. For example, in route 31-1, the time has been improved, but there is no improvement in the number of steps. The reason is that when there is a crowd, the number of steps increases and an alternative route is proposed that may increase the number of steps. In Fig. 13, impact numbers of steps taken with regard to distance and traffic has been very good. Therefore, the proposed algorithm performs more optimally for routes with diverse paths.

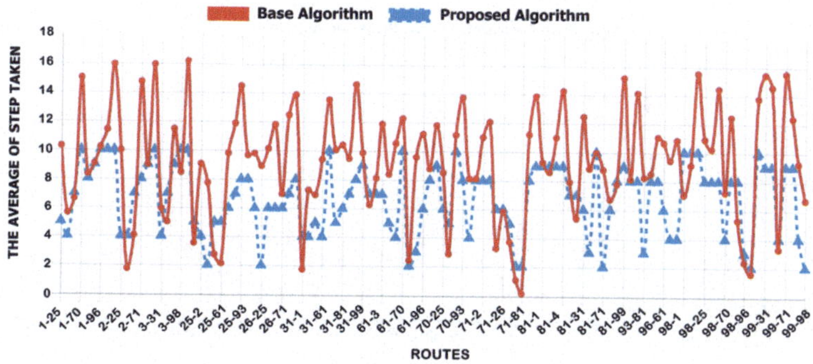

Fig. 13 Step taken

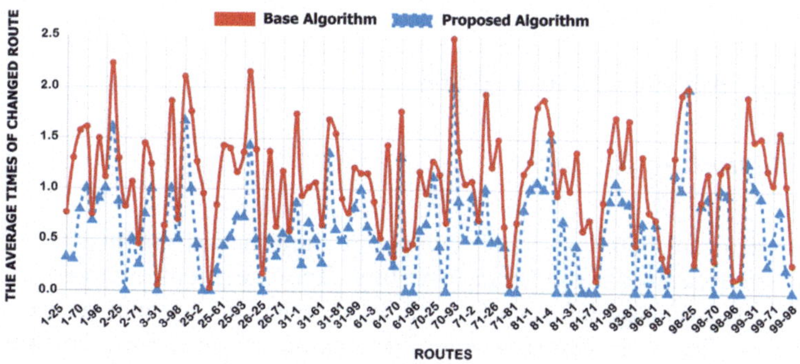

Fig. 14 Changed routes

6.4 *The Effect of Algorithm by Considering Obstacles*

Figures 14 and 15 show the congestion level in the environment controlled by the proposed algorithm. When facing an obstacle, the wall tracking algorithms is used. Moreover in times of overcrowding which is considered as mobile obstacle, different solutions are proposed for bypassing the obstacles and producing the optimal result. Our algorithm takes on-line real world data adaptation of the moment, history of the moments, and high-traffic times into consideration in order to make a better decision. In Fig. 14, as the user has 9 choices at any situation and one of them is selected based on the parameters and route traffic, solutions such as 81-4, 81-31, and 81-71 indicate different routes for controlling the obstacle. However, in solutions such as 1-70 or 26-25, the obstacle is in a route, which is controllable only from one side or two sides. Figure 15 shows the number of encounters with obstacles, and the numbers of obstructed routes leading to rerouting. In some route, there were obstructions; while in others, there was no obstacle. This indicates that it is possible to dynamically cover instances of encountering with obstructions via alternative routes using the proposed algorithm.

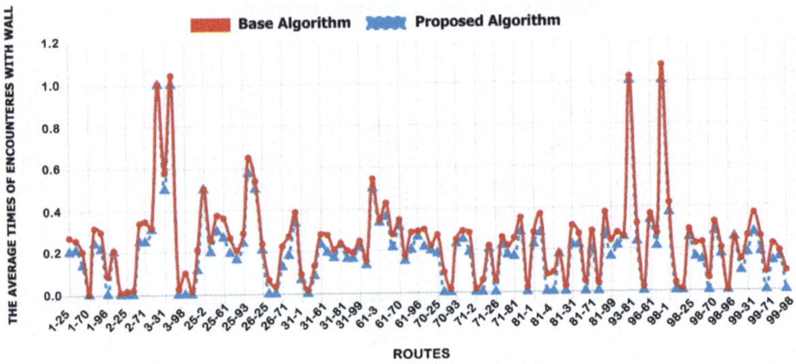

Fig. 15 Encounters with walls

7 Discussion and Future Work

7.1 Concluding Remarks

As we are all aware, in today's highly competitive information-oriented world, users' relationships are constantly changing over time. These relationship evolve and grow as businesses and users get to know each other better. In the input section of data mining, users life cycle indicates what information are available, and in the output section, the life cycle suggests what things are likely to be of interest and what decisions are supposed to be made. Data mining facilitates decision-making for organizational managers by conducting extensive data processing and decision-making processes through the extraction of valuable knowledge from the available real world data adaptation. Thus, the present study aims to address what valuable and useful information are extractable from the data stored upon users' requests and how this information can be used to inform timely and accurate decisions in order to ensure facilitation, in addition to satisfying the current users. Analyzing the real world data adaptation stored in databases reveals the services that are needed most and their limitations. The most important and applicable outcomes of data mining are reorganization of data for easier access, higher efficiency rates, and valid classification of data to obtain better results. Data mining techniques easily provide outcomes for each user based on their characteristics, so that organizations can rely on this information to predict how each class of users should behave in order to become satisfied by using patterns were collected and analyzed by using the embedded modules in mobile applications. According to the results, as time passes and more records are stored on the central server, previous experiences are used to suggest better routes, leading to shorter paths, less time consuming, fewer turns, and collisions with walls. Thus, it was concluded that the accuracy of routing with this method depends on the amount of collected data. More specifically, if the real world data adaptation collected exceeds over 50%, history of behavioral pattern usage can significantly improve INS and, thus, it lead to lower positioning errors and route improvement compared with the previous ones.

7.2 Discussion

In this study, the real environment conditions was simulated and indoor navigation was based on the scenes observed from simulation. To further accurate and realistic results, primary deployment by means of iBeacon or access point is indispensable. The significant challenge in deploying beacon-based navigation systems is the necessity of employing an efficient and costly beacon planning procedure to identify potential iBeacon placement locations [42]. Given the availability of infrastructure required for Wi-Fi-based positioning in public buildings, it is the most common deployed method in real world [43]. Moreover, Bluetooth Low Energy (BLE) beacons has good results on noise reduction and movement adaptation [44]. Using the developed mobile application and collecting data over a specified period, we will definitely achieve results that are more accurate. With regard to papers [45, 46], it is obvious that ML-based techniques are one of the best supplements for indoor navigation system's target achievements. Our proposed method is scalable in comparison with existing methods because of following reasons: First, initial deployment is not necessary for algorithm optimization experiment. Moreover, it leads to reduced congestion, less time, and best route suggestion. Although the computational cost of the proposed algorithm is high, it contributes to time and congestion reduction. In addition, depending on environmental changes, our system provisions the optimal route.

7.3 Future Works

For more accurate predictions, more effective fields must be integrated in data mining. For example, demographic features such as location, gender, age, education, occupation can affect data mining results and lead to improved outcomes. The field of data mining techniques and, of course, the use of a variety of other techniques for clustering and prioritizing users in areas with a large number of users, which is partly in the big data domain, is quite broad. Based on the use of data mining techniques, the following suggestions were made:

1. One of the interesting future direction is changing the placement of some features based on the collected information; for example, we can replace a nursing station or a doctor's office with other features to provide better services. Moreover, since most users visit the triage upon entering, we can place the triage closer to the entrance. We can also place frequently used stations in more accessible places,

enabling us to provide services for 2000 instead of 1000 user per day in the same (10 × 10) m space.
2. In this study, due to the special conditions of the simulations, time spent and steps taken factors were weighted equally. However, in real life, these factors can be weighted variedly based on the environment and also other factors such as energy.
3. Future studies may include the establishment of data mining analysis standards, such as data security, data retention technology for users' private information, data transparency, data performance measures, and the collaboration ability of the payment system.

8 Conclusion

In this study, an intelligent machine learning based indoor navigation system for IoT scenarios is presented where historical data were used to suggest better iteration in a smart scenario (e.g. hospital, airport). In addition, the proposed solution improves routing by decreasing the number of turns and encounter with obstacles. Our models for navigation has been formulated based on a heuristic algorithm by combining greedy and random forest algorithms for an efficient routing. Finally the performance of the proposed strategy is evaluated by simulation where the results demonstrated that the proposal scheme outperforms the existing methods in terms of traveled distance and elapsed time. Running the simulator on initial training data brought about a 25% improvement in comparison with the basic algorithm. In short, the proposed method provides better solution for intelligent indoor navigation and can be used in many smart cities scenarios where an optimal iteration is needed.

References

1. Abbasian Dehkordi, S., Farajzadeh, K., Rezazadeh, J., Farahbakhsh, R., Sandrasegaran, K., Abbasian Dehkordi, M.: A survey on data aggregation techniques in IoT sensor networks. Wirel. Netw. (2019)
2. Rezazadeh, J., Moradi, M., Sandrasegaran, K., Farahbakhsh, R.: Transmission power adjustment scheme for mobile beacon-assisted sensor localization. IEEE Trans. Ind. Inform. **15**(5), 2859–2869 (2019)
3. Shit, R.C., Sharma, S., Puthal, D., Zomaya, A.Y.: Location of things (LoT): a review and taxonomy of sensors localization in IoT infrastructure. IEEE Commun. Surv. Tutor. (2018)
4. Fortino, G., Savaglio, C., Palau, C.E., de Puga, J.S., Ganzha, M., Paprzycki, M., Montesinos, M., Liotta, A., Llop, M.: Towards Multi-layer Interoperability of Heterogeneous IoT Platforms: the INTER-IoT Approach, pp. 199–232. Springer International Publishing, Cham (2018). https://doi.org/10.1007/978-3-319-61300-0_10
5. Lin, K., Chen, M., Deng, J., Hassan, M.M., Fortino, G.: Enhanced fingerprinting and trajectory prediction for IoT localization in smart buildings. IEEE Trans. Autom. Sci. Eng. **13**(3), 1294–1307 (2016)

6. Mozaffari, N., Rezazadeh, J., Farahbakhsh, R., Yazdani, S., Sandrasegaran, K.: Practical fall detection based on IoT technologies: a survey. Internet of Things **8**, 100124 (2019). http://www.sciencedirect.com/science/article/pii/S2542660519302355
7. Agarwal, P., Gupta, A., Verma, G., Verma, H., Sharma, A., Banarwal, S.: Wireless monitoring and indoor navigation of a mobile robot using RFID. Nature Inspired Computing, pp. 83–90. Springer, Berlin (2018)
8. Rezazadeh, J., Sandrasegaran, K., Kong, X.: A location-based smart shopping system with IoT technology. In: 2018 IEEE 4th World Forum on Internet of Things (WF-IoT), pp. 748–753 (2018)
9. Nagarajan, S.G., Zhang, P., Nevat, I.: Geo-spatial location estimation for internet of things (IoT) networks with one-way time-of-arrival via stochastic censoring. IEEE Internet of Things J. **4**(1), 205–214 (2017)
10. Fortino, G., Trunfio, P.: Internet of Things Based on Smart Objects, Technology, Middleware and Applications. Springer, Berlin (2014)
11. Lashkari, B., Rezazadeh, J., Farahbakhsh, R., Sandrasegaran, K.: Crowdsourcing and sensing for indoor localization in IoT: a review. IEEE Sens. J. **19**(7), 2408–2434 (2019)
12. Mahida, P.T., Shahrestani, S., Cheung, H.: Localization techniques in indoor navigation system for visually impaired people. In: 2017 17th International Symposium on Communications and Information Technologies (ISCIT), pp. 1–6. IEEE (2017)
13. Deng, Z., Fu, X., Wang, H.: An IMU-aided body-shadowing error compensation method for indoor Bluetooth positioning. Sensors **18**(1), 304 (2018)
14. Yassin, A., Nasser, Y., Awad, M., Al-Dubai, A., Liu, R., Yuen, C., Raulefs, R., Aboutanios, E.: Recent advances in indoor localization: a survey on theoretical approaches and applications. IEEE Commun. Surv. Tutor. **19**(2), 1327–1346 (2016)
15. Khalajmehrabadi, A., Gatsis, N., Akopian, D.: Modern WLAN fingerprinting indoor positioning methods and deployment challenges. IEEE Commun. Surv. Tutor. **19**(3), 1974–2002 (2017)
16. Kao, C.-H., Hsiao, R.-S., Chen, T.-X., Chen, P.-S., Pan, M.-J.: A hybrid indoor positioning for asset tracking using Bluetooth low energy and Wi-Fi. In: 2017 IEEE International Conference on Consumer Electronics-Taiwan (ICCE-TW), pp. 63–64. IEEE (2017)
17. Zhang, R., Cui, Y., Claussen, H., Haas, H., Hanzo, L.: Anticipatory association for indoor visible light communications: light, follow me!. IEEE Trans. Wirel. Commun. **17**(4), 2499–2510 (2018)
18. Pathak, P.H., Feng, X., Hu, P., Mohapatra, P.: Visible light communication, networking, and sensing: a survey, potential and challenges. IEEE Commun. Surv. Tutor. **17**(4), 2047–2077 (2015)
19. Ding, W., Yang, F., Yang, H., Wang, J., Wang, X., Zhang, X., Song, J.: A hybrid power line and visible light communication system for indoor hospital applications. Comput. Ind. **68**, 170–178 (2015)
20. Rezazadeh, J., Subramanian, R., Sandrasegaran, K., Kong, X., Moradi, M., Khodamoradi, F.: Novel IBeacon placement for indoor positioning in IoT. IEEE Sens. J. **18**(24), 10 240–10 247 (2018)
21. Jeon, K.E., She, J., Soonsawad, P., Ng, P.C.: BLE beacons for internet of things applications: survey, challenges and opportunities. IEEE Internet of Things J. (2018)
22. Ahmetovic, D., Gleason, C., Kitani, K.M., Takagi, H., Asakawa, C.: NavCog: turn-by-turn smartphone navigation assistant for people with visual impairments or blindness. In: Proceedings of the 13th Web for all Conference, p. 9. ACM (2016)
23. S. F. Airport, LowViz guide-indoor navigation for the visually impaired. https://itunes.apple.com/us/app/lowviz-guide-indoor-navigation/id987917857?mt=8. Accessed 04 Sep 2016
24. Zheng, Y., Shen, G., Li, L., Zhao, C., Li, M., Zhao, F.: Travi-Navi: self-deployable indoor navigation system. IEEE/ACM Trans. Netw. **25**(5), 2655–2669 (2017)
25. Wang, Q., Guo, Y., Yang, L., Tian, M.: An indoor positioning system based on IBeacon. Transactions on Edutainment XIII, pp. 262–272. Springer (2017)

26. Rezazadeh, J., Moradi, M., Ismail, A.S.: Message-efficient localization in mobile wireless sensor networks. J. Commun. Comput. **9**(3), 340–344 (2012)
27. Vera, D., Marcillo, D., Pereira, A.: Blind Guide: anytime, anywhere solution for guiding blind people. In: World Conference on Information Systems and Technologies, pp. 353–363. Springer (2017)
28. Cecílio, J., Duarte, K., Furtado, P.: BlindeDroid: an information tracking system for real-time guiding of blind people. Procedia Comput. Sci. **52**, 113–120 (2015)
29. Meliones, A., Sampson, D.: Blind MuseumTourer: a system for self-guided tours in museums and blind indoor navigation. Technologies **6**(1), 4 (2018)
30. Michel, T., Genevès, P., Fourati, H., Layaïda, N.: Attitude estimation for indoor navigation and augmented reality with smartphones. Pervasive Mob. Comput. **46**, 96–121 (2018)
31. Sattarian, M., Rezazadeh, J., Farahbakhsh, R., Bagheri, A.: Indoor navigation systems based on data mining techniques in internet of things: a survey. Wirel. Netw. 1–18 (2018)
32. Zhu, Y., Mottaghi, R., Kolve, E., Lim, J.J., Gupta, A., Fei-Fei, L., Farhadi, A.: Target-driven visual navigation in indoor scenes using deep reinforcement learning. In: 2017 IEEE International Conference on Robotics and Automation (ICRA), pp. 3357–3364. IEEE (2017)
33. Li, L., Xu, Q., Chandrasekhar, V., Lim, J.-H., Tan, C., Mukawa, M.A.: A wearable virtual usher for vision-based cognitive indoor navigation. IEEE Trans. Cybern. **47**(4), 841–854 (2017)
34. Tariq, O.B., Lazarescu, M.T., Iqbal, J., Lavagno, L.: Performance of machine learning classifiers for indoor person localization with capacitive sensors. IEEE Access **5**, 12 913–12 926 (2017)
35. Zhang, W., Liu, K., Zhang, W., Zhang, Y., Gu, J.: Deep neural networks for wireless localization in indoor and outdoor environments. Neurocomputing **194**, 279–287 (2016)
36. Hirtle, S.C.: 8. wayfinding and orientation: cognitive aspects of human navigation. Handbook of Behavioral and Cognitive Geography, p. 141 (2018)
37. Schwarzkopf, S., Büchner, S.J., Hölscher, C., Konieczny, L.: Perspective tracking in the real world: Gaze angle analysis in a collaborative wayfinding task. Spat. Cogn. Comput. **17**(1–2), 143–162 (2017)
38. Anacta, V.J.A., Schwering, A., Li, R., Muenzer, S.: Orientation information in wayfinding instructions: evidences from human verbal and visual instructions. GeoJournal **82**(3), 567–583 (2017)
39. Kraemer, D.J., Schinazi, V.R., Cawkwell, P.B., Tekriwal, A., Epstein, R.A., Thompson-Schill, S.L.: Verbalizing, visualizing, and navigating: the effect of strategies on encoding a large-scale virtual environment. J. Exp. Psychol.: Learn. Memory Cogn. **43**(4), 611 (2017)
40. Shafique, U., Qaiser, H.: A comparative study of data mining process models (KDD, CRISP-DM and SEMMA). Int. J. Innov. Sci. Res. **12**(1), 217–222 (2014)
41. Gheisari, S., Meybodi, M.R.: LA-CWSN: a learning automata-based cognitive wireless sensor networks. Comput. Commun. **94**, 46–56 (2016)
42. Cheraghi, S.A., Namboodiri, V., Sinha, K.: IBeaconMap: automated indoor space representation for beacon-based wayfinding (2018). arXiv:1802.05735
43. Long, Z., Men, X., Niu, J., Zhou, X., Ma, K.: A Wi-Fi indoor positioning modeling based on location fingerprint and cluster analysis. In: International Conference on Computer Vision Systems, pp. 336–345. Springer (2017)
44. Gomes, A., Pinto, A., Soares, C., Torres, J.M., Sobral, P., Moreira, R.S.: Indoor location using Bluetooth low energy beacons. In: World Conference on Information Systems and Technologies, pp. 565–580. Springer (2018)
45. Sepulveda, G., Niebles, J.C., Soto, A.: A deep learning based behavioral approach to indoor autonomous navigation (2018). arXiv:1803.04119
46. Vandermeeren, S., Van de Velde, S., Bruneel, H., Steendam, H.: A feature ranking and selection algorithm for machine learning-based step counters. IEEE Sens. J. **18**(8), 3255–3265 (2018)

The manufacturer's authorised representative in the EU is Springer Nature Customer Service Centre GmbH, Europaplatz 3, 69115 Heidelberg, Germany. If you have any concerns regarding our products, please contact ProductSafety@springernature.com

Printed and bound by CPI Group (UK) Ltd, Croydon, CR0 4YY
22/09/2026
02221062-0003